# Energy Demand Challenges in Europe

Energy Demand Challenges in Europe

Frances Fahy · Gary Goggins ·
Charlotte Jensen
Editors

# Energy Demand
# Challenges in Europe

Implications for policy, planning and practice

*Editors*
Frances Fahy
School of Geography and Archaeology
and Ryan Institute
National University of Ireland Galway
Galway, Ireland

Gary Goggins
School of Geography and Archaeology
and Ryan Institute
National University of Ireland Galway
Galway, Ireland

Charlotte Jensen
Department of Planning
Aalborg University
Copenhagen, Denmark

ISBN 978-3-030-20338-2      ISBN 978-3-030-20339-9   (eBook)
https://doi.org/10.1007/978-3-030-20339-9

This Palgrave Pivot imprint is published by the registered company Springer Nature Switzerland AG
The registered company address is: Gewerbestrasse 11, 6330 Cham, Switzerland

# CONTENTS

# Notes on Contributors

**Desislava Asenova** works as an analyst at the Applied Research and Communications Fund, Bulgaria since 2013. She is part of the 'Science, Technology and Innovation Policy Programme' and has been involved in international research projects in various areas such as sustainable innovations, social innovations, public engagement in policymaking, and bi-regional STI cooperation and policy dialogue in the Black Sea region. She is currently working on projects related to energy systems and energy conscious behaviour of households.

**Julia Backhaus** is a doctoral student at the International Centre for Integrated assessment and Sustainable development (ICIS) at Maastricht University. Working on sustainable energy and consumption topics at the Energy research Centre of the Netherlands and at Maastricht University over the past ten years, Backhaus has published on subjects ranging from social innovation and sustainable food practices to energy policies and a refuelling infrastructure for hydrogen-fuelled cars. Her Ph.D. research investigates actors' assumptions about how change towards sustainability will come or can be brought about. She is a founding and still active member of the Sustainable Consumption Research and Action Initiative (SCORAI) Europe.

**Laure Dobigny** is a Postdoctoral Researcher at the Institute of Sociological Research at the University of Geneva, specialised in the socio-anthropology of energy. Her research interests centre around the social construction and social dimension of energy, especially linked to the energy transition (e.g. energy communities, renewable energy use and representations, sociotechnical systems, the relationship between production and consumption in the daily energy use, household consumption and self-consumption, among others).

**Frances Fahy** is Senior Lecturer in Geography at the National University of Ireland (NUI) Galway and Lead Coordinator of the European H2020-funded ENERGISE project. Her research interests are in environmental planning and sustainability.

**Audley Genus** is Professor of Innovation and Technology Management at Kingston University. He holds a Ph.D. in technology policy and innovation from Aston University. Audley's current research concerns the contribution of users and diverse organisations to innovation in and growth of the renewable energy sector, and problems of institutional change relating to sustainable consumption and production. His latest book is *Social Innovation and Sustainable Consumption*, co-edited with several colleagues from the Horizon 2020-funded ENERGISE project and others and published in 2018 by Routledge.

**Gary Goggins** is ENERGISE project manager at NUI Galway. He holds a Ph.D. in Environmental Sociology and Sustainability Studies, M.A. in Community Development and a degree in Business Studies. His research interests include sustainable consumption and sociotechnical change.

**Eoin Grealis** is an environmental economist and current member of scientific staff at the Department of Geography at LMU Munich. He received a B.A. Int. (Law and Economics) and an M.A. in Economic and Environmental Modelling from NUI Galway. After working as an energy policy analyst with the Irish Wind Energy Association, Eoin conducted Ph.D. research at NUI Maynooth on Irish agricultural emissions. He subsequently held a Postdoctoral position with the Socio-Economic Marine Research Unit (SEMRU) at NUI Galway where he helped develop the Bio-Economy Input-Output model (BIO) for Ireland. He has also published on the spatial distribution of farm incomes, farm viability and policy reform.

**Marko Hajdinjak** is a senior analyst at Applied Research and Communications Fund, Bulgaria. He has 20 years of experience in participation in and management of international research projects. His main research interests include social-cultural aspects of domestic energy consumption and energy efficiency; social innovations; responsibility in research and innovation; socio-economic impacts of science and research; and safe and responsible use of digital technologies among children and young people.

**Eimear Heaslip** completed her B.Arch. (HONS) in Architecture at University College Dublin, graduating in 2007. In 2012 Eimear completed an M.Sc. in Environmental systems in GMIT and received the GMIT 'President's Forty Year Scholarship' to undertake her transdisciplinary Ph.D., which fused social scientific and engineering techniques to develop a technical energy plan with the case study community in Inis Oírr. Eimear is currently working as a Postdoctoral Researcher on the ENERGISE (European Network for Research, Good Practice and Innovation for Sustainable Energy) project and as an Assistant Lecturer in Environmental Planning based in the School of Geography at NUI Galway.

**Eva Heiskanen** is Professor of sustainable consumption at University of Helsinki, Finland. Research interests: the adoption and impacts of new technology in society, sustainable consumption policies, evaluation and learning from interventions and experimentation.

**Marfuga Iskandarova** is a Postdoctoral Researcher at the Small Business Research Centre, at Kingston Business School. Her doctoral research at the University of Exeter focused on development of renewable energy technologies in the UK. Marfuga's research interests include innovation, energy policy, and sustainable consumption. She is currently working on the European Commission Horizon 2020-funded 'ENERGISE' project investigating socio-economic, cultural and political aspects of energy consumption.

**Charlotte Jensen** is Assistant Professor at Aalborg University, Copenhagen. She has an M.Sc. in innovation and environmental management from the Technical University in Denmark, and holds a Ph.D. in sociology of technology, consumption and transition from Aalborg University. Charlotte's research focuses on exploring links between consumption and production patterns, and environmental and social implications thereof. Several of her publications deal with social scientific understandings of (energy) demand, hereunder the wider implications of such approaches for energy policy and (sustainable) change.

**Senja Laakso (Ph.D.)** is a Postdoctoral Researcher at University of Helsinki, Finland. Research interests: sustainable consumption, social practices and transition paths towards sustainable food and energy systems.

**Kaisa Matschoss (Ph.D., Adjunct Professor)** works as a university researcher at University of Helsinki, Finland. Research interests: sustainable use of energy, experimentation and innovation in energy transition.

**Annika-Kathrin Musch** studied Geography at the University Eichstätt-Ingolstadt, Germany from 2008 until 2014 and deepened her understanding of economic geography and regional planning with study visits at the University of Nottingham, UK, and the Matej Bel University Banská Bystrica, Slovakia. She is a key member of the LMU-based INOLA research team, investigating innovations for regional sustainable energy management using transdisciplinary participatory methods and Living Lab techniques. Annika's Ph.D. research focuses on planning, processing and evaluating participation in transdisciplinary research, with a view to identifying participatory practices that have empowering and transformative potential.

**Patrick Naef** Anthropologist and geographer, is a senior research associate at the University of Geneva, looking at the transformation of cities, energy efficiency and the critical approaches of resilience. He explored these topics principally in Colombia, Bosnia-Herzegovina, Croatia and Switzerland. He is also collaborating with several community organisations active in cultural and urban development. Patrick Naef obtained a Ph.D. in geography at the University of Geneva in 2014. He was a visiting scholar at the Department of Anthropology of UC-Berkeley from 2014 to 2016.

**Maj-Britt Quitzau** is Associate Professor at Aalborg University, Copenhagen. She has an M.Sc. in Environmental Engineering and did her Ph.D. on sociotechnical transformations of bathroom practices. Her research is mainly related to sustainable transitions with special focus on innovation processes in the building sector and urban planning. She has a special interest in studying how strategic planning practices succeed in impacting urban development, both with regards to technologies and daily practices.

**Henrike Rau** is Professor of Social Geography and Sustainability Research at LMU Munich. Her extensive research work to date has addressed the topics of (un)sustainable consumption, mobility across the life course and mobility cultures and urban development. She has also made internationally recognised contributions to the conceptual and methodological advancement of social-scientific sustainability research, especially regarding the critical application of the 'new mobilities paradigm' and the investigation of household consumption practices. As a member of the Sustainable Consumption Research and Action Initiative (SCORAI Europe) steering committee, she is actively shaping the sustainable consumption research landscape in Europe and beyond.

**Inge Røpke** is Professor of Ecological Economics at Aalborg University, Copenhagen. She has published widely on ecological economics, consumption and environment, consumption in a practice theory perspective, energy use and information technology in everyday life as well as consumers' role within the smart grid. Her recent research has been focused on ecological macroeconomics and issues of finance and sustainability. She was the 2014 recipient of the Nicholas Georgescu-Roegen Award for Unconventional Thinking and the 2018 recipient of the Kenneth E. Boulding Memorial Award.

**Marlyne Sahakian** is Assistant Professor of Sociology at the University of Geneva, where she brings a sociological lens to consumption studies in relation to sustainability themes. Her research interest is in understanding everyday social practices, in relation to environmental concerns and social equity. Her inter- and transdisciplinary research relates to food consumption and energy usage in urban spaces, as well as societal wellbeing, and she writes regularly on these themes. She co-founded SCORAI Europe in 2012—a network in the field of sustainable consumption research and action, and is a board member of the European Sociological Association's Consumption Research Network.

**Tomislav Tkalec** is the energy programme coordinator in Focus Association for Sustainable Development. His work is mainly focused on topics of energy transition, energy efficiency, renewables, energy poverty, community energy projects, energy policies, and political ecology. He is experienced in working on FP7, IEE, HORIZON2020 and LIFE projects, and is coordinator of the Sustainable Energy Working Group in the national network of environmental organisations Plan B. His other tasks include advocacy and policy work, educational and awareness raising activities, and communication with media. His thesis Political aspects of decentralisation of electricity generation earned him a Ph.D. in Environment Protection at University of Ljubljana.

**Edina Vadovics (M.Sc., M.Ed., M.Phil.)** is research director of GreenDependent Institute and president of GreenDependent Association in Hungary, both with the mission to promote and research sustainable lifestyles, and facilitating dialogue between research and practice. Ms. Vadovics manages GreenDependent's work in international research projects, and campaigns. While her research focuses on sustainable communities and lifestyles, she is also involved in action projects. Earlier, she worked in corporate sustainability management, and taught related courses at various universities. She worked as an expert to the EEA and UNEP, and contributed to their key publications. She is a member of the SCORAI Europe steering committee.

**Lidija Živčič** is the senior expert of Focus. Her work covers topics of sustainable development, climate, energy, transport and degrowth. She has over 15 years of experience in managing projects at the level of Slovenia and EU. Lidija has extensive experience in education programs, awareness raising actions and policy campaigns. She is connected to a broad network of experts, decision-makers and campaigners in Slovenia and Europe. Completed an M.Sc. course in Environmental Science and Policy at Central European University in Budapest in 2001. Finished Ph.D. degree at Biotechnical faculty, University of Ljubljana, in 2012 with dissertation on raising awareness on climate change in Slovenia.

# LIST OF FIGURES

# LIST OF TABLES

# CHAPTER 1

# An Introduction to Energy Demand Challenges in Europe

*Frances Fahy and Gary Goggins*

**Abstract** This opening chapter calls for greater attention to energy demand challenges in Europe. It argues that many obstacles and opportunities in achieving the so-called energy transition are social and cultural in nature and require interdisciplinary solutions that go beyond efficiency approaches. We provide an overview of the ENERGISE project that aims to achieve greater understanding of the social and cultural influences on household energy use, and to develop appropriate responses and recommendations for policy-makers, practitioners and future academic research. The chapter concludes with a brief summary of the structure of this book, including an introductory overview of how energy demand challenges are understood, and how this relates to the types of solutions that are proposed in each of the ten European countries studied.

F. Fahy (✉) · G. Goggins
School of Geography and Archaeology and Ryan Institute,
National University of Ireland Galway, Galway, Ireland
e-mail: frances.fahy@nuigalway.ie

G. Goggins
e-mail: gary.goggins@nuigalway.ie

F. Fahy et al. (eds.), *Energy Demand Challenges in Europe*,
https://doi.org/10.1007/978-3-030-20339-9_1

1

**Keywords** Energy policy · Europe · Horizon 2020 · Interdisciplinary · ENERGISE

## Exploring Energy Demand Challenges in Europe

Significant challenges lie ahead regarding Europe's transition towards a decarbonised energy system that meets the economic and social needs of its citizens. Heretofore, scientific research and public policy in the field of household energy use has primarily focused on promoting energy efficiency through changes in technology and individual behaviour (Labanca and Bertoldi 2018). However, these approaches have been ineffective in bringing about the aggregated reductions in carbon emissions that are necessary to meet climate targets. They may even be counterintuitive if they reinforce unsustainable routines and habits that engage energy-related services (Shove 2018; Hargreaves et al. 2018).

The performance of more efficient technologies is often dependent on how they are used by householders, if at all. Moreover, short-term efficiency gains may be wiped out by increasing overall consumption in order to reach newly perceived levels of comfort, convenience and standards. This increase in consumption manifests, for example, through social pressure to consume in line with the latest 'trends' for bigger houses, bigger cars, the latest technologies and appliances, and so forth. At the same time, many households experience energy poverty and are unable to meet their energy needs, such as providing adequate heat for their homes. Addressing these concerns requires multi-dimensional approaches and highlights the crucial role of consumption in multi-scalar decarbonisation efforts across Europe, an issue requiring much greater attention from scientists and policy-makers than before.

This calls for a broadening of discussions around energy use to include such fundamental and deep-rooted questions as 'How much of what is enough?' Of course, finding consensus on a clearly subjective issue is never going to be easy, or indeed possible. But, by framing the energy challenge in this way, we are compelled to be reflexive in considering how collective conventions around energy use evolve, and how can they be better aligned with sustainable lifestyles. Opening up these increasingly pertinent discussions to wider debate also implies a greater emphasis on citizen engagement to respond to related social and cultural challenges through more participatory processes. Undertaking these democratic exercises can help us to better comprehend societal norms and routines that greatly determine our patterns of energy use as well as our ability to change those patterns. By understanding and accounting for particular householder needs and other contextual conditions of consumption, policy-makers and practitioners can tailor sustainable energy responses accordingly.

This book gathers together contributions from prominent social scientists researching in the energy field all across Europe. While the authors hail from diverse disciplinary backgrounds, they all recognise that cultural- and systemic change is a key ingredient in successful energy transitions. The book offers unique and often fascinating insights into the socio-material similarities and differences in energy policies, energy infrastructures and energy demands across 10 European countries. The collection provides invaluable accounts of the diverse contexts within which individuals, households and communities engage in everyday energy-related practices, and the policies and practices that underpin ongoing efforts towards sustainable transformation.

## THE ENERGISE PROJECT AS CONTEXT

This book is a key output from the ENERGISE project. ENERGISE is an innovative pan-European research initiative to achieve a greater scientific understanding of the social and cultural influences on energy consumption. Funded under the EU Horizon 2020 programme for three years (2016–2019), ENERGISE develops, tests and assesses options for a bottom-up transformation of energy use in households and communities across Europe.

Energy use can be fruitfully understood as engaging in a set of practices that incorporate elements of meaning, skills and material conditions and that are embedded in wider social, political and institutional contexts (Shove et al. 2012). This practice-oriented perspective is central to the ENERGISE project for a number of reasons. First, a practice perspective departs from individualistic views of energy choices and behaviour that have unduly limited past research on energy demand and its transformation. Instead, the focus shifts towards an understanding of energy behaviour as collectively shared and culturally mediated. Second, an explicit focus on energy-related practices promotes cutting-edge social-scientific and interdisciplinary energy research that covers both social and material aspects of energy use.

In particular, ENERGISE investigates the energy-related practices of households and communities and their impacts on society and the environment. This is achieved through a living lab approach, where researchers work with households in a real-world setting and in the context within which their energy use takes place. ENERGISE Living Labs (ELLs) build on interactions between various stakeholders as well as lessons drawn from previous initiatives aimed at reducing household energy use, or what we term sustainable energy consumption initiatives (SECIs). In ENERGISE, SECIs are defined as activities that deal with reducing energy-related $CO_2$ emissions from households, and enables active participation from households. This can either be in terms of (1) reducing the actual energy consumption, (2) reducing the emissions intensity of energy consumption (e.g. by substituting fossil fuels with renewable energy sources). Sustainable energy initiatives are considered to be socio-technical because they attempt to change the material arrangements and the cultures, norms and conventions that determine collective energy use and related impacts. Examples of good practice SECIs that have informed the design and implementation of ELLs are provided for each of the central chapters in the book. The results of the ELLs themselves are presented elsewhere.

## REMIT OF THE BOOK

The following chapter will introduce the concept of 'problem framings' related to the way energy consumption challenges are understood and the impact for the type of solutions that emerge. Based on the review of SECIs already carried out across Europe as part of the ENERGISE project, it is evident that many different perspectives on sustainable energy use exist and problems are addressed in a number of different ways.

Energy use is typically interpreted as a matter of individual choices and preferences that can be changed independently of the context within which consumption occurs, and often through technological optimisation or incremental behaviour change. This approach, however, often fails to translate into significant changes in energy consumption patterns and energy demands. In contrast, other more integrated approaches treat energy consumption challenges as a matter of wider societal, cultural and institutional dynamics, which suggest that changes in energy demand will come about only if the entire range of elements that underpin energy use are considered. These alternative approaches set out to obtain quite different results; some suggest qualitative changes in habits and routinised activities, while others set sufficiency-based targets that seek absolute reductions in energy use (or increases for those in energy poverty) and set minimum and maximum limits to consumption. Illustrated with examples, Chapter 2 will introduce how these different approaches or problem framings can be characterised, and what the different framings mean for the objectives, methods and assumptions made in sustainable energy consumption initiatives.

The central chapters in this collection provide insights into trends in energy transitions and what this means for the future of energy demand in Europe. The inclusion of ten chapters, each focusing on a different European country, offers a broad yet accessible overview of the diverse energy systems currently in place, albeit in a rapidly changing context. What is particularly evident is that no one system is free from contestation over how to best meet present and future energy needs. Every country is unique in the energy-related challenges they face, with each having different institutional structures and capacity to deal with problems and to capitalise on opportunities. At the same time, all countries share common concerns around issues such as energy security, affordability and sustainability, and all have signed up to international climate change agreements that call for dramatic reductions in carbon emissions.

Another interesting observation is how energy policy is evolving across the countries under study. Emerging trends include the changing role of citizens in the energy transition, more generally portrayed as shifting from 'passive' to 'active' consumers, but manifesting in a number of different forms from 'prosumers' to 'energy communities'. Householders are also envisaged to play a central role as the shift toward 'smart' grid solutions and systems intensifies. Driving these changes requires robust policy responses and the engagement of actors at all levels of society including government bodies, researchers, businesses, NGOs and community groups.

A recent phenomenon that facilitates greater engagement in decision-making has been a political shift across much of Europe toward multilevel governance. This has facilitated greater societal engagement with the sustainable energy agenda and provided new avenues for non-state actors to play a significant role in developing responses.

The expansion of the spatial and political boundaries in which responses occur is reflected in the increasing range of actors, sites, configurations and mechanisms through which sustainable energy is being addressed, as demonstrated in the case studies provided in this book. The final chapter within this collection reflects on the key strands emerging from the material presented and considers individual and collective opportunities for sustainable energy transitions. In comparing and contrasting energy-related problem framings and social, material and institutional make-up across Europe, the concluding chapter unpacks the energy challenges facing Europe. It clearly shows that policies for energy demand reduction have to carefully consider and address the differences in cultural, material and institutional constitutions of energy demand and energy systems, locally, regionally, nationally and cross-nationally.

## References

Hargreaves, T., Wilson, C., & Hauxwell-Baldwin, R. (2018). Learning to live in a smart home. *Building Research & Information, 46*(1), 127–139.

Labanca, N., & Bertoldi, P. (2018). Beyond energy efficiency and individual behaviours: Policy insights from social practice theories. *Energy Policy, 115*, 494–502.

Shove, E. (2018). What is wrong with energy efficiency? *Building Research & Information, 46*(7), 779–789.

Shove, E., Pantzar, M., & Watson, M. (2012). *The dynamics of social practice: Everyday life and how it changes*. Los Angeles: Sage.

# CHAPTER 2

# Framing the Sustainable Energy Challenge and Implications for Solutions

*Charlotte Jensen, Inge Røpke, Gary Goggins and Frances Fahy*

**Abstract** Sustainable consumption policies often rely on ecological modernisation rationality, where the focus is usually on making current consumption patterns more sustainable in such a way that status quo (ideas about the quality of life and growth) is not challenged. As a result, sustainable energy policies tend to black box the demand-side, often resulting in abstracting efficiency strategies from the social

C. Jensen (✉) · I. Røpke
Department of Planning, Aalborg University, Copenhagen, Denmark
e-mail: cjensen@plan.aau.dk

I. Røpke
e-mail: ir@plan.aau.dk

G. Goggins · F. Fahy
School of Geography and Archaeology and Ryan Institute,
National University of Ireland Galway, Galway, Ireland
e-mail: gary.goggins@nuigalway.ie

F. Fahy
e-mail: frances.fahy@nuigalway.ie

F. Fahy et al. (eds.), *Energy Demand Challenges in Europe*,
https://doi.org/10.1007/978-3-030-20339-9_2

organisation within which the strategies and resulting solutions unfold. Rebound effects and other unintended consequences often happen as a result of this type of efficiency strategies. This chapter introduces alternative problem framings that may offer a way to mitigate rebound effects by addressing and challenging a wider set of socio-material, cultural and institutional aspects of energy demand.

**Keywords** Energy demand · Problem framings · Energy policy · Sustainable consumption · Transformation

## INTRODUCTION

Over the past few decades, consumer-oriented environmental policies have proliferated as political and public interest and concern for environmental issues have increased significantly (Christensen et al. 2007). As a result, it is not uncommon that citizens as consumers are assigned responsibility for sustainable development. This in itself may not be a problem, but it is important to highlight how responsibility is assigned, in what way, and what it means for the type of development that emerges and is emphasised as a result. As sustainable consumption policies often rely on ecological modernisation rationality, these policies often centre on making current consumption patterns more sustainable in such a way that new business opportunities can emerge and 'quality' of life is not challenged (Sedlacko et al. 2014). As a result energy policies tend to focus on making existing behaviours more sustainable; a focus that often ends up abstracting efficiency strategies from the social organisation within which the strategies and resulting solutions unfold (Labanca and Bertoldi 2018).

In this chapter, we open with a discussion of the type of common *energy problem framings* that appear embedded in most consumer-oriented (sustainable) energy policies, and what these problem framings imply for the type of results obtained. We then discuss alternative energy problem framings that may be beneficial to implement in consumer-oriented (sustainable) energy policies. Approaches for example that might mitigate potential rebound effects or other unintended consequences that often are the result of the dominant types of efficiency strategies. We facilitate this discussion by introducing the ENERGISE Problem

Framing Typology, that highlights different dimensions and aspects of the problem framings most commonly used ('technology-orientated' and 'individual behaviour-oriented') as well as of problem framings that are much less used ('everyday life oriented' and 'systems oriented'). The latter we argue can offer insights into how social, cultural and institutionalised aspects of energy consumption can be investigated and challenged as part of the efforts of reducing energy consumption levels.

## COMMON ENERGY PROBLEM FRAMINGS

It is widely acknowledged that environmental and climate change policies often build on dominant paradigms of economics and psychology. This means that the theories of change embedded in many policies pave the way for the assumption that climate change problems are a result of individual actions, which can be changed by addressing attitudes, behaviours and choice (Shove 2010). This is often operationalised in strategies that attempt to shift people's choices away from unsustainable or inefficient products towards more sustainable or efficient products, primarily through information and the promotion of (energy) efficient products (Spurling et al. 2013). As Labanca and Bertoldi (2018: 496) state, such strategies often assume that solutions can be 'surgically removed and replaced by other solutions, seamlessly entering the social tissue where they are installed, without causing any change but reduction in energy inputs'.

Although technological innovation can bring about significant efficiency potentials, these may only be realised if appropriate economic instruments are applied simultaneously, so that gains from efficiency strategies are not just directed towards other unsustainable areas (Christensen et al. 2007; Shove 2017). This means that if energy efficiency strategies are applied without addressing and potentially disrupting systems of interacting, unsustainable practices that generate high levels of energy demand, the energy demand problem is not addressed but potentially only shifted, instigating a possible rebound effect.

As Southerton and Welch (2018) highlight, required reductions in consumption-related emissions cannot be achieved through marginal lifestyle changes and technical efficiencies. While the environmental impact of economic outputs has been reduced in advanced economies, the relationship between growth in per capita income and growth in per capita GHG emissions continues. This finding is supported by Bjørn et al. (2018), who find that if levels in consumption-related demand are not lowered, technological development will not deliver the requirements to meet the climate goals set out in the Paris Agreement. Problem framings that understand energy use and consumption as a result of technological efficiency and incremental lifestyle (behaviour) changes thus have seemingly limited potential to achieve the needed fundamental changes in energy demand levels.

According to the ENERGISE typology of energy problem framings (Jensen et al. 2017), two main types of problem framings are often employed within current sustainable energy consumption initiatives (SECIs), technology-orientated or behaviour-orientated. Consequently, a majority of initiatives either take technological development or changes in individual behaviours as the main drivers of change.

What does this look like in practice? A typical example of a SECI underpinned by a technological problem framing would be if energy use related to, for example, heating is solely (or at least primarily) understood to be a matter of optimising heating systems. Optimisation could also include a focus on providing (technical) labelling for heating systems so that the 'consumer' can easily navigate between different settings in terms of energy efficiency. Such SECIs, however, do not explicitly challenge the extent and duration for which people heat their homes, nor would they fundamentally challenge any notions related to maximum or minimum temperatures. SECIs within this framing category would therefore not explicitly challenge what is understood to be appropriate levels of indoor comfort in different contexts and situations.

A general illustration of a SECI underpinned by a behaviour change type of problem framing might go a bit further than relying on energy efficiency labelling of products. Often this expanded approach puts emphasis on providing more information about why it is good for the consumer to choose an efficient heating system, or why the consumer should turn down their thermostat. Information provided may focus on monetary incentives or it may address ecological consequences of not choosing the most energy efficient option. It does however not challenge socially shared norms around heating, or what it means to feel 'comfortable'.

## ALTERNATIVE ENERGY PROBLEM FRAMINGS

As several researchers (e.g. Spurling et al. 2013; Southerton and Welch 2018; Genus et al. 2018) suggest, it may instead be beneficial to establish new problem framings that lead to altogether fundamentally different modes of governance and policies that disrupt unsustainable practices, or reconfigure links between practices. In this way, it is not energy consumption in itself that is targeted (and made efficient) but rather, it is what energy is for that is scrutinised and challenged (Shove and Walker 2014). Related energy problem framings pose altogether different questions about what needs to be changed, why and how. The resulting solution space(s) would be broader (and more complex) than for dominant energy problem framings, and would presuppose a reflexive mode of governance. According to the ENERGISE typology of energy problem framings, two types of alternative energy problem framings follow along these lines of rationale, however they are only rarely utilised in existing SECIs. One of these problem framings understands everyday life situations as being the central point of departure for change potentials, and another understands complex interactions between multiple actors, systems and practices as being the point of departure for potential change.

In contrast to the common energy problem framings discussed earlier, SECIs that seek to address heating related energy use, and which are underpinned by an everyday life situations problem framing, would approach energy use altogether differently. In these cases, the situations of everyday life that have an implication for the way, as well as the frequency and extent to which, people heat their homes would thus be the 'unit' of interventions. Here, SECIs might target routines and ideas related to how, why and when people heat their homes in different types of situations. This could be in terms of challenging ideas about norms and comfort that can vary across situations, for example when receiving guests. Solutions within this problem framing might include heating people instead of spaces, providing low-energy alternatives to space heating such as wearing more clothes, using blankets or rearranging furniture. In that way, SECIs underpinned by this problem framing may address understandings of comfort and material aspects of heating, and may employ a notion of sufficiency rather than efficiency (see Sahakian et al. 2019).

In the context of SECIs that adopt a problem framing that presumes a broader interaction between multiple actors, systems and practices to be central to change, these initiatives would target energy use related to heating as a matter of challenging current ideas about comfort and routines related to heating homes. In addition, they would argue for (or ideally even enable) political and legislative changes in terms of how energy for heating is produced and distributed, so that consumers and communities benefit from a low-carbon transition. SECIs underpinned by this type of problem framing could also challenge norms and standards for increasingly bigger homes (increasing number of square meters per person), as the savings gained by efficient heating systems are often offset by houses getting increasingly bigger (Christensen et al. 2007). These SECIs target a wider range of actors, challenge existing ways of organising everyday life around more sustainable systems of production and consumption, as well as building community and institutional networks for sustainable transformation of buildings. Eco-communities often resemble such attempts, by socially and materially organising different ways for producing energy as well as spaces for sharing particular types of activities, sometimes resulting in smaller private homes, which enable people to engage in alternative heating practices.

## How Different Types of Problem Framings Generate Different Objectives, Targets and Outputs

The four types of problem framings discussed above are summarised below in terms of implied objectives, methods of intervention, consumption areas targeted, types of outputs and types of change. The two first types of problem framings (Changes in Technology and Changes in Individuals Behaviours) belong to what we here describe as *common* energy policy problem framings. The last two types of problem framings (Changes in Everyday Life Situations and Changes in Complex Interactions) belong to what is in this chapter termed *alternative* energy problem framings. The typology categories and highlighted dimensions of change summarised below are based on a large-scale review of 1000+ recent and current European SECIs (Jensen et al. 2017).[1]

---

[1] For an overview of SECIs classifications, please visit our database http://energise-project.eu/projects

**Table 2.1**  Objectives, methods of interventions and types of outputs across problem framings

| Dimension of change / Typology category | Objectives | Methods of intervention | Consumption area targeted | Type of output | Type of change |
|---|---|---|---|---|---|
| Changes in technology | Focus is on providing householders with opportunities to make technological improvements in their homes | Information, sometimes monetary incentives and legislation, some forms of experimentation | Often non-specific energy use, and if specific it has a technological focus (e.g. heating- or lighting systems) | Emissions and energy saved | Technological, mostly at appliance and building level |
| Changes in individual behaviour | Focus is on raising awareness about climate change and energy use. Often focus on energy efficiency | Information, campaigns, training and some forms of peer-to-peer learning | Often unspecific; general electricity use | Energy efficiency in (use of) appliances, sometimes raised awareness, often rebound effects | Often non-specific; focus on behaviour as a matter of energy awareness. Sometimes including specific measures of nudging |
| Changes in everyday life situations | Changing consumption patterns by targeting need/configuration for instance by substituting practices (for instance away from private driving towards public transportation) | Often a broad mix, usually an element of community and more experimental forms of deliberation, such as collaboration or 'living lab' approaches | Often specific in relation to particular domains in or related to the home (heating, use of water, cooking). Can be more general, but then often with a focus of using less (sufficiency) | Often expressed through changes in consumption patterns related to energy use, less explicitly about energy use. Sometimes expressed through energy demand reductions | Often expressed as changes in practices related to household-specific situations (cooking and in relation to heating). Often about changes in mobility patterns |

(continued)

**Table 2.1**  (continued)

| Dimension of change / Typology category | Objectives | Methods of intervention | Consumption area targeted | Type of output | Type of change |
|---|---|---|---|---|---|
| Changes in complex interactions | Often expressed as changes across 'supply and demand'; can be development of new relations between renewable energy system providers and users; new ways of living, new ways of building; development of professional as well as everyday life practices. Often related to specific locations | Often a broad mix, very often involving experimentation and community based action | Often several areas targeted, and often across professional practices (e.g. building, banking, planning) and household related practices (with several forms of energy use) | Often a mix of better performance of buildings and changes in energy demand. Often resulting in new relationships across professional and everyday life domains. Often involving new ways of planning | Often expressed as changes in several types of consumption patterns and as changes in how practices interlock |

*Source* Adapted from Jensen et al. (forthcoming)

Table 2.1 provides an overview which can provide a guide or insight into what type of outputs might result from particular types of objectives, and related methods of intervention.

## ENERGISE REVIEW OF 1000+ SECIs CATEGORISED ACCORDING TO THE PROBLEM FRAMING TYPOLOGY

The ENERGISE review of 1000+European SECIs has resulted in an overview of the share of SECIs that are underpinned by each of the four typology categories presented above.[2] At least 75% of the SECIs are underpinned by common energy problem framings, whereas less than 25% of SECIs seem underpinned by problem framings that challenge underlying dynamics of (and reasons for) energy demand. Interestingly, SECIs underpinned by alternative energy problem framings are also primarily small-scale and local, reaching far less people and actors than SECIs underpinned by classic energy problem framings. This is problematic, as current energy problem framings tend to prioritise (abstract) efficiency strategies, which may (1) obscure longer-term trends in demand and societal shifts in what energy is for and (2) reproduce specific, potentially unsustainable, understandings of 'service', including perceived standards for comfort and convenience (Shove 2017) (Table 2.2).

Drawing on the results of this extensive European review, each of the following chapters in this collection concludes by showcasing a good practice example of a European sustainable energy consumption initiative underpinned by alternative problem framings that take either everyday life situations or broader systemic complex interactions between several practices and systems to be the target of intervention. The case studies are practical 'real world' examples which are intended to serve as inspiration for anyone who would like to know more about how initiatives underpinned by alternative problem framings can be designed and operationalised.

---

[2]For more information about the ENERGISE review and classification of SECIs, please consult Jensen et al. (2018) and Jensen et al. (2017).

**Table 2.2**    Overview of ENERGISE classification of SECIs

|  | No. of initiatives | % of total initiatives | Sub-national (e.g. local; regional) | National/ cross-national |
|---|---|---|---|---|
| Sustainable energy consumption initiatives (SECIs)—total | 1067 | 100 | 398 | 669 |
| Changes in technology | 284 | 26.6 | 101 | 183 |
| Changes in individual behaviour | 513 | 48 | 153 | 360 |
| Changes in everyday life situations | 123 | 11.5 | 56 | 67 |
| Changes in complex interactions | 147 | 13.8 | 88 | 59 |

## REFERENCES

Bjørn, A., Hauschild, M., Kabins, S., Jensen, C., Schmidt, J., & Birkved, M., et al. (2018). Pursuing necessary reductions in embedded GHG emissions of developed economies: Will efficiency improvements and changes in consumption get us there? *Global Environmental Change, 52*, 314–324.

Christensen, T., Godskesen, M., Gram-Hanssen, K., Quitzau, M., & Røpke, I. (2007). Greening the Danes: Experience with consumption and environment policies. *Journal of Consumer Policy, 30*, 91–116.

Genus, A., Fahy, F., Goggins, G., Iskandarova, M., & Laakso, S. (2018). Imaginaries and practices: Learning from 'ENERGISE' about the integration of social sciences with the EU Energy Union. In *Advancing energy policy* (pp. 131–144). Cham: Palgrave Pivot.

Jensen, C., Goggins, G., & Fahy, F. (2017). *Construction of typologies of sustainable energy consumption initiatives*. ENERGISE—European Network for Research, Good Practice and Innovation for Sustainable Energy, D2.4.

Jensen, C., Goggins, G., Fahy, F., Grealis, E., Vadovics, E., Genus, A., et al. (2018). Towards a practice-theoretical classification of sustainable energy consumption initiatives: Insights from social scientific energy research in 30 European countries. *Energy Research and Social Science, 45*, 297–306.

Jensen, C., Goggins, G., Røpke, I., & Fahy, F. (forthcoming). *Achieving sustainability transitions in residential energy consumption across Europe: Do problem framings within existing initiatives match current and future needs?*

Labanca, N., & Bertoldi, P. (2018). Beyond energy efficiency and individual behaviours: Policy insights from social practice theories. *Energy Policy, 115,* 494–502.

Sahakian, M., Naef, P., Jensen, C., Goggins, G., & Fahy, F. (2019). Challenging conventions towards energy sufficiency: Ruptures in laundry and heating routines in Europe. In *ECEEE Summer Study 2019 Proceedings.*

Sedlacko, M., Martinuzzi, A., Røpke, I., Videira, N., & Antunes, P. (2014). Participatory systems mapping for sustainable consumption: Discussion of a method promoting systemic insights. *Ecological Economics, 106,* 33–43.

Shove, E. (2010). Beyond the ABC: Climate change policy and theories of social change. *Journal of Environment and Planning, 42,* 1273–1285.

Shove, E. (2017). What is wrong with energy efficiency? *Building Research & Information, 46,* 1–11.

Shove, E., & Walker, G. (2014). What is energy for? Social practice and energy demand. *Theory, Culture and Society, 31*(5), 41–58.

Southerton, D., & Welch, D. (2018). *Transitions for sustainable consumption after the Paris agreement.* The Stanley Foundation. Available at https://www.stanleyfoundation.org/publications/pab/SustainableConsPAB1118.pdf.

Spurling, N., McMeekin, A., Shove, E., Southerton, D., & Welch, D. (2013). *Interventions in practice: Re-framing policy approaches to consumer behaviour.* Sustainable Practices Research Group. Available at http://eprints.lancs.ac.uk/85608/.

# CHAPTER 3

# The Impact of German Energy Policy on Household Energy Use

*Eoin Grealis, Annika-Kathrin Musch and Henrike Rau*

**Abstract** This chapter reviews current energy policy and civil society efforts to achieve the targets set out for Germany's *Energiewende* (Energy Transition), with a specific focus on their impact on household energy use. The existing energy governance structure, conflicting energy policy commitments, and the emergence of public resistance to renewable infrastructure are identified as significant challenges for national-level policy. At the household level, the dominance of efficiency and smart choice solutions and the pressure to maintain traditional patterns of consumption are identified as key limiting factors in an effort to deliver real reductions in household energy use.

**Keywords** Energy policy · Household energy use · Energy infrastructure · Germany · Energy transition

E. Grealis (✉) · A.-K. Musch · H. Rau
LMU Munich, Munich, Germany

A.-K. Musch
e-mail: a.musch@lmu.de

H. Rau
e-mail: Henrike.rau@lmu.de

© The Author(s) 2019
F. Fahy et al. (eds.), *Energy Demand Challenges in Europe*,
https://doi.org/10.1007/978-3-030-20339-9_3

## INTRODUCTION

Germany has committed to a number of targets in order to achieve a successful *Energiewende* (energy turn), the transition to a sustainable energy system. These include 60% of total energy demand being supplied from renewable resources and a total greenhouse gas emission (GHG) reduction target of 80–95% of 1990 levels by 2050 (BMWi 2015). While the country has recently made progress in some areas, particularly with the deployment of renewable energy capacity in the electricity sector, current forecasts suggest that Germany will miss its interim GHG emissions target of a 40% reduction on 1990 levels by 2020. Instead, the Federal Ministry projects a reduction of 32% over the same period. Additionally, recent policy commitments to a faster phasing out of nuclear power have limited nuclear's capacity as a non-carbon-intensive 'bridging technology' (albeit one that presents other sustainability risks) until such time when technological and infrastructural solutions are in place to ensure the stability of supply with large-scale renewable deployment. In order to achieve its stated emissions targets, Germany must reduce annual GHG emissions at a consistent rate of 3.5% per annum on current levels for the next 30 years. Considering that the 30-year historical average is a reduction of 1.2%, and the 10-year average just 0.8%, this presents a significant challenge (EEA 2019), which must be addressed by several sectors, including households.

## SOCIO-MATERIAL DYNAMICS OF HOUSEHOLD ENERGY USE IN GERMANY

Households account for approximately a quarter of total energy demand in Germany, ranking third behind the industrial (47%) and services sectors (26%) in 2017 (BDEW 2018). Total residential energy demand is dominated by space (68%) and water (14%) heating requirements, with the remaining energy use accounted for by lighting and appliances (8%), cooking (6%) and other uses (4%) (Eurostat 2018a). Household energy consumption is still dominated by fossil fuels, including oil- and gas-fired heating systems (gas 37%, oil 22%), with electricity (21%[1]),

---

[1] Of 20.8%, 12.9% was generated from fossil fuels sources in 2017 with the remaining 7.9% generated from renewable sources (Fraunhofer Institute 2018).

renewables (11%), derived heat (8%) and solid fuels (1%) providing the rest (Eurostat 2018b). Germany ranks 6th out of the EU28 in terms of per capita electricity consumption, with the International Energy Agency reporting an average consumption of just over 7000 kWh per annum in 2014 (IEA 2014).

Dwelling location, type, size, tenure and household composition are significant factors in the determination of household energy use. At 40%, Germany has the third highest proportion of single-person households in the EU (Eurostat 2019a), which translates into a comparatively higher average per capita living space. With a predominantly clustered pattern of settlement, 77% of the 82.9m population live in either an urban or predominantly urban location, with just 23% classified as living in rural areas (Eurostat 2019b). Consequently, apartment and other shared dwellings serve the majority (approx. 60%) of German households, with the remaining residences consisting of a mixture of detached (26%) and semi-detached and terraced housing (Eurostat 2019b). The high instance of apartment living has meant that shared private spaces and services are widespread and culturally acceptable. For example, shared laundering facilities in the basements of apartment complexes are a common occurrence while utility costs associated with common areas are often divided among residents.

With regard to tenure, lifelong tenancy is a common and widely accepted cultural phenomenon in Germany. Germany has the highest rate of household tenancy (49%) and the lowest rate of home ownership (51%) in the EU (although higher than Switzerland) (Eurostat 2019c). This can be attributed to a mix of both historical redevelopment (inclusion of the private building sector in post-war social housing schemes) and comparatively strong tenancy rights. However, high tenancy rates have the potential to reduce the market for sustainable investments if tenants do not receive a share of the benefits, both in terms of energy saving measures (e.g. retrofitting) and renewable generation (e.g. solar). This has been recognised under recent amendments to the Renewable Energy Act (2014), which ensures that tenants and landlords both receive a share of return from sustainable energy investments through a subsidy paid for domestic electricity production (BMJV 2018).

## ENERGY POLICY IN GERMANY

Energy policy at the federal level in Germany is dominated by supply-side and infrastructural legislative measures, largely influenced by both its implementation obligations under EU directives[2] and national policy documents. Technical regulations derive primarily from the Energy Industry Act governing grid network charges, transmission, reserve, access and default supply, while regulations governing the transition to less carbon-intensive energy supply stem primarily from the Renewable Energy Sources Act (BMJV 2018). The latter broadly promotes the advancement of technological and market solutions that will enable Germany to reach a target of 80% of power generation from renewable resources by 2050.

Recently, Germany has made significant progress on the penetration of renewables, particularly in the electricity sector, with renewables accounting for 38% of net public power supply in 2017 (Fraunhofer Institute 2018). However, the accident at the Fukushima nuclear power plant in 2011, and subsequent political developments in Germany (particularly in the state of Baden-Württemberg[3]), have had a significant impact on Germany's nuclear energy policy, with the Federal government removing nuclear power as a defined 'bridging technology' towards achieving the *Energiewende* and legislating for a decommissioning timeline for most nuclear power plants by 2022 (BMJV 2018). As a result, increased levels of renewables may now result in a corresponding increase in the use of fossil fuels for electricity production (Renn and Marshall 2016). Consequently, attention among *Energiewende* campaigners and advocates has shifted towards decommissioning coal-related technology and infrastructure, as exemplified by the *Ende Gelände* movement that seeks to stop the use of coal.

In addition to the technical barriers to renewable expansion, there is also mounting public resistance to *Energiewende* projects, most notably in relation to wind farm developments and the construction and

---

[2] Flowing from both the EU Climate and Energy Package (governing the 20/20/20 targets) and the European Climate and Energy Framework (outlining targets for 2030) policy frameworks.

[3] In 2011 the Green Party won the State elections in Baden-Württemberg, partly because of its ability to capitalise on the Fukushima incident to promote proposals for a rapid 'energy turn' towards renewables.

upgrading of power lines. In 2014, the German government, respond-
ing to increasing public resistance to the implementation of local energy
transition projects and the required upgrading and expansion of the electricity grid, agreed to slow down the expansion of renewable energy projects and limit further expansion to 'development corridors' as well as
revising the aims of the Renewable Energy Act (2014). While the primary policy focus had up to this point been on the accelerated decarbonisation of energy used to create electricity, recent developments are
beginning to shift attention towards demand-side policies, together with
associated ordinances such as the law on the eco-design of energy-related
products (BMJV 2017).

While overarching policies are developed at a Federal level, much of the
practical actions required to deliver a successful *Energiewende* fall on local
municipalities and civil society. The highly devolved nature of local administration means that local municipalities are primarily responsible for either
taking direct action (in terms of their own energy use/supply) or indirect action such as providing the necessary conditions for private/semi-
private sustainable investments or supporting civil society actions aimed at
lowering energy consumption. The capacity and extent of municipalities'
engagement in such actions is a result of the available resources (budg-
etary and physical), the extent of devolved administrative competence,
and the level of political prioritisation, which may differ considerably
both across and within each State/Bundesland. As a consequence, the
*Energiewende* is likely to progress at different rates across Germany. In
fact, in some instances past policy both at the Federal and State levels has
clearly undermined renewable energy development with overly prescrip-
tive development restrictions, resulting in an effective ban of particular
types of generation in certain areas (Naßmacher and Naßmacher 2007).
For example, the controversial '10H' regulation, introduced in Bavaria in
2014, requires any installed turbine to be a minimum distance of 10 times
the turbine height from any residential building (Bayerische Bauordnung
2018). This effect of this regulation has been to substantially restrict the
siting options for wind energy infrastructure.

The opportunity of private individuals and communities to participate
directly in the transition by investing in renewable energy projects has
been central to the successful increase in the proportion of renewable
energy used in electricity generation over the last ten years. It has been
estimated that 46% of installed renewable capacity was owned by farmers
and private citizens in 2012 (Borchert and Wettengel 2018). However,

changes to the Renewable Energy Act appear to have caused a drop off in co-operative investment regulations, with a transition away from feed-in-tariff supports to bid or auction systems (Morris 2014). Such systems can be problematic for citizen co-operatives, which typically only plan to realise a single project and cannot therefore split the risk of a lost auction in contrast to large-scale commercial developers who can bid for several projects at once (Amalang 2016).

Although more recent signals at the Federal level clearly favour a more centralised *Energiewende*, the German energy transition also involves a decentralisation of energy production and the emergence of new actor networks in so-called energy regions (Gailing and Röhring 2016). Between the level of municipalities and the federal level, the formation of collaborative network-based governance is considered a determining factor for the success of regional energy transitions (Gailing 2018). An example for a bottom-up regional multi-level governance project can be found in the Oberland region in Southern Bavaria. Here, *Energiewende Oberland* (EWO), a civic foundation for energy transition founded in 2005, has been identified as the decisive actor responsible for institutionalising the energy transition and for building effective networks and governance structures regarding the energy sector in the region (Von Streit and Bothe, in review).

Yet despite the emergence of active regional networks (such as in the Oberland region), efforts towards a decentralised energy transition continue to meet unfavourable political and regulatory conditions set at Federal and State levels that favour a centralised *Energiewende*.

## Trends in National Household Energy Campaigns in Germany

National household energy campaigns have primarily focused on, and prioritised technical solutions, with the primary future vision for a successful *Energiewende* reliant on improved technical innovation, greater energy efficiency, passive/carbon positive housing, improved energy transmission, and high-tech grid management in order to enable greater proliferation of renewables (BMWi 2018). This trend is also evident in campaigns aimed at changing energy-related behaviour. Here, the focus is firmly on encouraging individuals to make 'smarter' consumer choices in terms of more efficient lighting, heating systems and household appliances, with reduction of use strategies featuring much less prominently. A recent analysis of 60 sustainable energy consumption initiatives (SECIs) in Germany

**Table 3.1**  Problem framings of 60 sustainable energy consumption initiatives from Germany

| Problem framing | No. of initiatives |
| --- | --- |
| Changes in individuals' behaviour | 27 |
| Changes in technology | 21 |
| Changes in complex interactions | 8 |
| Changes in everyday life situations | 4 |

*Source* Jensen et al. (2018)

(Jensen et al. 2018) closely mirrored trends in national policy, with the majority of initiatives aimed at changing technology and consumer behaviour. The continued absence of sufficiency-based strategies (e.g. the re-evaluation of necessary consumption) is a notable omission given the growing evidence of the inability of efficiency based strategies to fully achieve anticipated reductions (Druckman et al. 2011) (Table 3.1).

There is also significant stratification when it comes to particular targeted areas of energy use, with many initiatives targeting one particular aspect such as retrofitting, information campaigns targeting purchasing behaviour, saving-potential analysis, and energy or emission saving competitions. The large number of SECIs profiled demonstrates a general commitment to improving environmental awareness and the willingness to contribute to energy saving and climate protection; however, the emphasis on saving (energy and/or money) and other participatory incentives reveals that there is a current expectation that SECIs should provide 'added value' for participants.

Certain 'basics' or cultural norms around consumption would appear to be less palatable for discussion or negotiation (e.g. car ownership, holiday travel, meat consumption) and have not been targeted specifically in SECIs related to energy initiatives. This was also one of the findings of the Energiesuffizienz Project, a collaborative research initiative funded by the Federal Ministry for Education and Research.

## Case Study: Energiesuffizienz Project

The research project 'Energiesuffizienz' (energy sufficiency) was undertaken from 2013 to 2016 and was tasked with identifying the driving factors and dynamics for the expansion of energy-related 'needs' and how they could be addressed, with a view to achieving real quantitative reductions in the size and use of devices, the substitution of technical equipment in households, and the adjustment or reassessment of technical services delivered by appliances to utilities and desired by users (i.e. smart grid services) (Brischke et al. 2016). The approach concentrated on three elements: households, appliances, and urban infrastructure and services in municipalities. A criteria-based analysis was conducted that examined action and measurement options for energy sufficiency in the distinct areas of living and building, as well as individual barriers and framework conditions that influence or hinder the implementation of energy sufficiency. Based on this theoretical framework, empirical studies were carried out which employed transdisciplinary methods.

### Investigative Approach

Households represented the core subjects of investigation in the project. In a representative survey of 601 households, the research team enquired as to how energy-sufficiency practices are currently perceived and evaluated, what sufficiency practices are already employed, and whether and to what extent additional sufficiency practices may be accepted in the future. For example, in the area of living space, participants were asked to rate the adequacy of their current living space ranging from 'much too small' to 'far too big'. Additionally, there were interviews with several actors at the municipal level to analyse existing measures and approaches to improve energy sufficiency. The research team used neighbourhood labs that drew on five local communities of practice (youth group, local co-op, a group of degrowth activists, senior citizens club and a Christian seniors group). The research team presented cultural probes to get to know the participants and their performances of practices also within the group, and held co-creation workshops to counter conflicts with handling sufficiency strategies.

### Framing the Energy Challenge

Brischke et al. (2016) argue that the existing policy measures that foster the *Energiewende* in Germany concentrate primarily on improving energy efficiency, and that they ignore energy-sufficiency strategies to a large extent. They note that while energy efficiency in many sectors has been consistently improved, total energy use has remained stable. They further note that efficiency is only one factor of total energy use, and point to the fact that the technical characteristics (size, features, etc.), use patterns and total number of appliances have a significant bearing on overall energy use. The authors further argue that energy efficiency improvements are being eaten up by higher levels of consumption and/or rising expectations of comfort. Consequently, the authors argue that as there are technical and economic limits on energy efficiency, energy sufficiency will be an important aspect in the energy transition.

In the Energiesuffizienz study, energy was framed as a consumer product that in and of itself held little interest for households in their day-to-day lives, and that energy sufficiency measures should be developed in such a way that consumers become aware of which needs and wishes are important for a high quality of life (and, conversely, which are not). Departing from traditional policy interventions that try to reduce the energy intensity of practices without questioning established consumption levels, the approach challenges this prevailing 'optimization orthodoxy' by seeking to lower demand and reduce energy use.

### Outcomes/Outputs

Brischke et al. (2016) found (among other observations) that energy sufficiency practices can play a large role in efforts to reduce energy use. These sufficiency practices are already present in many households and are regarded as normal. Moreover, they are not correlated with financial endowment and can be implemented irrespective of incomes. The Energiesuffizienz Project also found that sufficiency practices are less acceptable in leisure activities than during core household duties, and that playing on people's environmental consciousness or 'guilt tripping' people into action are not necessarily helpful strategies for promoting sufficiency. Sufficiency thinking is largely incompatible with the current status quo, i.e. a growth-based economic system and the dominance of efficient technological fixes in sustainability thinking, policy and practice. This cannot be easily reconciled with the (more or less) radical change agenda that underpins much sufficiency thinking, i.e. that less may in fact be more (Grealis and Rau 2018).

Through open innovation workshops, the project design guide also provided detailed and specific eco-design sufficiency recommendations relating to the *reduction* (e.g. display and adjustability of cooling temperature, instead of an abstract scale in refrigerators and freezers), *substitution* (supporting the change in practices and routines towards energy and resource conservation through innovative design of the appliances, e.g. washing with low temperatures, measured laundry dosing), and *adjustment* of appliances (e.g. equipment should be designed such that functions and features only consume energy, when they are in use).

## Conclusion

While early progress has been made in the area of renewable electricity generation, Germany faces significant future challenges in this area. Similarly, more sustained efforts are needed to achieve real reductions in respect of other forms of energy consumption. While Federal policy and legislation provide the overarching targets, the devolved and partially fragmented nature of energy governance in Germany and its impact on bottom-up actions by both local municipalities and civil society mean the future rate of change for the energy transition is very uncertain. Environmental consciousness and public support for the *Energiewende* are generally rather high. However, transition measures which fail to include citizens and local energy co-operatives may slow progress as they are more likely to meet resistance from below, especially concerning both grid development and installation of renewable energy technology. Additionally, some aspects of German culture may clash with efforts to achieve real reductions in energy consumption as part of the *Energiewende*. In particular, Germany's prevailing 'car culture' frequently combines with a historically influential automobile industry to undermine or weaken efforts towards a *Verkehrswende* (sustainable transport turn) that links closely with sustainable energy goals discussed in this chapter. The recent heated public debate in Germany concerning proposals for a speed limit of 130 km per hour on all German highways exemplifies this conflict. Overall, this overview has shown the need for future policy and change programmes that address systems of inter-locking energy-related practices (e.g. mobility, space and water heat-ing, domestic appliance use), as opposed to focusing solely on changing Germany's energy supply system from the top-down.

## REFERENCES

Amalang, S. (2016). *Germany's energy transition revamp stirs controversy over speed, participation.* Clean Energy Wire (Online News publication). Accessed 10 January 2018 from https://www.cleanenergywire.org/dossiers/reform-renewable-energy-act#controversial.

Bayerische Bauordnung. (2018). Bavarian Building Regulations (BayBO) as amended on 14 August 2007 (GVBl. P. 588, BayRS 2132-1-B), most recently amended by § 1 of the Act of 10 July 2018 (GVBl. P. 523).

Borchert, L., & Wettengel, J. (2018). *Citizens' participation in the Energiewende.* Clean Energy Wire (Online News publication). Accessed 10 January 2018 from https://www.cleanenergywire.org/factsheets/citizens-participation-energiewende.

Brischke, L. A., Leuser, L., Duscha, M., Thomas, S., Thema, J., Spitzner, M., et al. (2016). *Energiesuffizienz – Strategien und Instrumente für eine technis-che, systemische und kulturelle Transformation zur nachhaltigen Begrenzungdes Energiebedarfs im Konsumfeld Bauen/Wohnen.* Heidelberg, Berlin, and Wuppertal: IFEU, Institut für Energie- und Umweltforschung Heidelberg.

Druckman, A., Chitnis, M., Sorrell, S., & Jackson, T. (2011). Missing carbon reductions? Exploring rebound and backfire effects in UK households. *Energy Policy, 39*(6), 3572–3581.

EEA. (2019). *National Inventory Report of German greenhouse gas inventory 1990–2017.* Accessed 21 January 2019 from http://cdr.eionet.europa.eu/de/eu/mmr/art07_inventory/ghg_inventory/envxd4xlg/2019-01-15_EU_NIR_2019.pdf.

Eurostat. (2018a). *Share of final energy consumption in the residential sector by type of end-use, 2016.* Accessed 9 January 2019 from https://ec.europa.eu/eurostat/statistics-explained/index.php?title=Energy_consumption_in_households#Data_sources_and_availabilityEurostat.

Eurostat. (2018b). *Share of fuels in the final energy consumption in the residential sector, 2016 (nrg_110a).* Updated 31 May 2018.

Eurostat. (2019a). *Distribution of households by household type from 2003 onwards—EU-SILC survey (ilc_lvph02).* Updated 19 January 2019.

Eurostat. (2019b). *Distribution of population by degree of urbanisation, dwelling type and income group—EU-SILC survey (ilc_lvho01).* Updated 19 January 2019.

Eurostat. (2019c). *Distribution of population by tenure status, type of household and income group—EU-SILC survey (ilc_lvho02).* Updated 19 January 2019.

Federal Association of Energy and Water Industry (BDEW). (2018). *Energy demand by demand group 2017.* https://www.bdew.de/media/documents/Nettostromverbrauch-nach-Verbrauchergruppen-2017_online_o_jaehrlich_Ki_27042018.pdf.

Federal Ministry for Economic Affairs and Energy (BMWi). (2015). *The energy of the future: Fourth "energy transition" Monitoring Report—Summary.* Berlin, Germany. Accessed on 17 January 2019.

Federal Ministry for Economic Affairs and Energy (BMWi). (2018). *Unsere Energiewende: sicher, sauber, bezahlbar.* Accessed 27 January 2019 from https://www.bmwi.de/Redaktion/EN/Dossier/energy-transition.html.

Federal Ministry for Justice and Consumer Protection (BMJV). (2017). Ordinance implementing the law on the ecodesign of energy-related products (EVPGV) amended 18 January 2017 (BGBl. I S. 85).

Federal Ministry for Justice and Consumer Protection (BMJV). (2018). Renewable Energy Sources Act amended 17 December 2018.

Fraunhofer Institute. (2018). *Power generation in Germany—Assessment of 2017.* Accessed 27 January 2019 from www.energy-charts.de.

Gailing, L. (2018). Die räumliche Governance der Energiewende: Eine Systematisierung der relevanten Governance-Formen. In O. Kühne & F. Weber (Eds.), *Bausteine der Energiewende* (pp. 75–90). Wiesbaden: Springer Fachmedien Wiesbaden.

Gailing, L., & Röhring, A. (2016). Is it all about collaborative governance? Alternative ways of understanding the success of energy regions. *Utilities Policy, 41,* 237–245.

Grealis, E., & Rau, H. (2018). *Exploring the (in)compatibilities of efficiency and sufficiency thinking in the context of efforts to reduce domestic energy use.* Third International Conference of the Sustainable Consumption Research and Action Initiative (SCORAI): Sustainable Consumption: Fostering Good Practices and Confronting the Challenges of the 21st Century. Copenhagen.

IEA. (2014). *Electric power consumption (kWh per capita).* Accessed 27 January 2019 from http://data.worldbank.org/indicator/EG.USE.ELEC.KH.PC?end=2014&locations=DE-IE-FR-FI-GB-DK-LV-LT-HR-CZ-IT-SE-ES-SI-SK-RO-PL-PT-MT-NL-LU-HU-GR-EE-CY-BG-AT-BE&start=2014&view=bar.

Jensen, C. L., Goggins, G., Fahy, F., Grealis, E., Vadovics, E., Genus, A., & Rau, H. (2018). Towards a practice-theoretical classification of sustainable energy consumption initiatives: Insights from social scientific energy research in 30 European countries. *Energy Research & Social Science, 45,* 297–306.

Morris, C. (2014). *Bundestag adopts new rules for renewables.* Energie Transition: The Global Energiewende. Accessed 20 January 2019 from https://energytransition.org/2014/07/bundestag-adopts-new-renewable-energy-act/.

Naßmacher, H., & Naßmacher, K.-H. (2007). *Kommunalpolitik in Deutschland.* Wiesbaden: Springer.

Renewable Energy Act. (2014). (EEG) (Federal Law Gazette I p. 1066), last amended by Article 1 of the Act of 17 December 2018 (BGBl. I P. 2549).

Renn, O., & Marshall, J. P. (2016). Coal, nuclear and renewable energy policies in Germany: From the 1950s to the "Energiewende". *Energy Policy, 99,* 224–232.

Von Streit, A., & Bothe, J. (in review). Institutional work and spatial practices within regional transitions towards renewable energies: Experiences from southern Germany.

CHAPTER 4

# The Role of Households in Danish Energy Policy: Visions and Contradictions

*Inge Røpke, Charlotte Jensen and Maj-Britt Quitzau*

**Abstract** This chapter outlines the transformation of the Danish energy system from the oil crises in the 1970s to the present challenges. Energy policies have ensured a successful implementation of district heating based on combined heat and power and high penetration of wind power in the electricity system, but also substantial dependence on the use of biomass. The transformations have concentrated mainly on the supply side, where the involvement of households has been somewhat scattered. Turning to the future challenge of decarbonization, more focus on the demand side is needed, including energy savings and the electrification of mobility and heating. In this process, households may need to be involved differently. An innovative example of multi-actor engagement in energy renovation of private homes is presented.

I. Røpke (✉) · C. Jensen · M.-B. Quitzau
Department of Planning, Aalborg University, Copenhagen, Denmark
e-mail: ir@plan.aau.dk

C. Jensen
e-mail: cjensen@plan.aau.dk

M.-B. Quitzau
e-mail: quitzau@plan.aau.dk

© The Author(s) 2019
F. Fahy et al. (eds.), *Energy Demand Challenges in Europe*,
https://doi.org/10.1007/978-3-030-20339-9_4

35

**Keywords** Denmark · Danish energy system · Transformation ·
Energy savings · Role of households in energy transformation

## INTRODUCTION

This chapter outlines developments and tendencies in Danish national
energy policy, contextualized in brief descriptions of the Danish energy
system as well as current energy reduction approaches.

The chapter is divided into four overall sections, beginning with a
short, historical illustration of how and why the current Danish energy
system has come to be. Subsequently, current trends in Danish energy
policy are presented and discussed in relation to how the role of households is portrayed in current and future strategies for low carbon transitions. Danish national energy policy and approaches tend to utilize
mainstream understandings, or problem framings, of energy demand,
wherefore strategies and approaches for energy reductions usually take
on technocratic or consumer-behaviour oriented perspectives on change.
There are, however, a few examples of Danish sustainable energy consumption initiatives that take a broader perspective of change, and bring
forward the need for practice-based, systemic changes in order to obtain
low carbon transitions. The chapter will provide a short introduction to
one such initiative, followed by an overall conclusion.

## THE DANISH ENERGY SYSTEM

The Danish energy system has been through continuous transformations
since the oil crises in the 1970s. Presently (the late 2010s), the system is
characterized by a relatively high penetration of wind power in the electricity system, a large share of district heating based on combined heat
and power (CHP), a high degree of self-sufficiency in energy, and considerable use of (imported) biomass.

The first steps towards a modern utilization of wind power were taken
by pioneers and popular movements in the wake of the oil crises, but
the endeavour played a limited role in the first decades and met with
much resistance from incumbent interests. Instead, electricity companies
focused on replacing oil with coal, and tried to promote the introduction of nuclear power, but this was blocked by a combination of factors,
including public resistance. When global warming began to attract public

attention in the late 1980s, politicians forced electricity companies to invest in wind turbines. Gradually, wind power gained an increasing role in the system, comprising 43% of annual average electricity demand in 2017 (Energistatistik 2017).

The expansion of district heating and CHP was another result of the oil crises, which encouraged considerable heat planning efforts aimed at decentralization. Previously, nearly all electricity was produced by large central power stations, which were gradually converted to CHP to provide district heating to larger cities. In addition, existing local district heating plants were converted to also generate electricity, and the number of decentralized CHP plants increased considerably, standing at about 400 today (Dansk Fjernvarme 2018).

When the oil crises emerged, fossil fuel extraction from the Danish part of the North Sea was in its infancy. The government decided to establish a system that could make use of the related natural gas to replace oil in residential heating. Two collective pipe-based systems were thus established: direct provision of natural gas to households (and other sectors) and district heating based on CHP. Heat planning stipulated which areas should be supplied in which way. By and large, both the electricity system and CHP plants were collectively owned by consumers or municipalities until about the year 2000, after which the system underwent liberalization and partial privatization (Hvelplund 2007). Combined with legislation that allowed municipalities to commit consumers to connect to the collective systems, this form of ownership enabled a remarkable transformation to a more rational energy utilization. While about 25% of households were connected to district heating in 1975, today about 65% of households use district heating, while 15% are heated with natural gas (Wistoft et al. 1992: 204; Energistyrelsen 2018). In an EU context, a similar penetration of district heating occurs only in the Baltic countries.[1]

Resource extraction from the North Sea also provided oil, and, as production increased, the dependency on imports fell. The degree of self-sufficiency in total energy use grew from 5% in 1980 to 52% in 1990, and in 1997 Denmark became energy self-sufficient (Dietrich and Morthorst 2016).

---

[1] http://www.euroheat.org/wp-content/uploads/2016/03/2015-Country-by-country-Statistics-Overview.pdf.

Peaking in 2004, the degree of self-sufficiency reached 155%, but since then the production of oil and gas has fallen, and the degree of self-sufficiency fell to 85% in 2017 (Energistatistik 2017).

Security of supply was the main concern in the wake of the oil crises and encouraged conversion from oil to coal in power plants. When climate concerns later intensified, the phase-out of coal emerged on the political agenda. Local CHP plants adopted the use of a variety of fuels including wood pellets, waste, straw, natural gas and biogas, and more recently, large power plants increasingly converted from coal to biomass. Since 2000, the use of biomass has more than doubled, with imports accounting for 43% of biomass used in 2016 (Klimarådet 2018). With the inclusion of biomass, renewable energy constituted 33% of Danish energy consumption in 2017, up from 6% in 1990 (Energistatistik 2017).

When corrected for trade in electricity and weather, total Danish energy consumption was 5.7% lower in 2017 than in 1990. In terms of $CO_2$ emissions, Danish emissions fell by 38% from 1990 to 2017 (Energistatistik 2017). However, the Danish $CO_2$ emissions per capita are still a little above the EU average in 2016.[2] Part of the explanation why Denmark lacks behind some rich countries, such as Sweden and the UK, is that Denmark neither has hydropower, nor nuclear power.

## CURRENT TRENDS IN ENERGY POLICY

As the outline above illustrates, the transformations of the Danish energy system have concentrated mainly on the supply side. But some of the changes involved the cooperation of households, who changed their heating installations from oil burners to district heating installations and natural gas boilers. The oil crises also led to other initiatives on the demand side, including campaigns aimed at persuading people to lower the temperature in dwellings and turn off lights. Subsidies were provided for thermal insulation and double-glazing, and building regulations were tightened. Compulsory energy labelling of appliances was introduced, and campaigns to shut off standby mode on appliances were implemented (Christensen et al. 2007). In spite of all these initiatives,

---

[2] https://ec.europa.eu/eurostat/tgm/table.do?tab=table&init=1&language=en&pcode=t2020_rd300&plugin=1.

energy consumption has been relatively stagnant. Both population and living standards have increased, implying increased car ownership, more square metres per person, more appliances, more leisure travel, etc. In addition—and not included in these accounts—part of the energy consumption related to improved living standards has been outsourced to other countries as part of globalization.

The top priority of Danish energy policy is currently to reduce carbon emissions. All parties in Parliament agreed in 2018 on the goal that Denmark should be carbon neutral in 2050. This means that carbon can only be emitted if the emissions are compensated by a similar uptake in soil or forests or through technologies that capture and store carbon. This Energy Agreement includes measures to reduce emissions from the energy sector and energy-intensive industrial plants (covered by the EU Emissions Trading System, ETS), for instance, the establishment of offshore wind farms and other investments in wind, solar power and biogas.[3] However, little is included to reduce emissions from the sectors that are not covered by ETS, such as transport, agriculture and buildings. These issues are due to be dealt with in separate agreements, but the challenge is tough due to substantial political disagreements.

In the long term, the extensive use of biomass for energy purposes is problematic because the land use competes with food production and the protection of biodiversity. Presently, the proposed solution is electrification of mobility and heating, based on electricity from wind and solar power. Households are encouraged to buy electric cars and to replace remaining oil burners with heat pumps, and the local CHP or heat plants are encouraged to invest in large heat pumps instead of using biomass. While some steps towards electrification have been taken, much remains to be done. One incentive for electrification has been the reduction of electricity tariffs, but this measure also has the unintended effect of reducing an incentive for energy savings. To avoid this rebound, electric cars and heat pumps would have to be promoted more effectively by targeted measures. If electrification succeeds, the need for wind and solar power will increase more than corresponding to the planned investments. Furthermore, the demand for increased capacity is amplified, as Denmark presently attracts large data centres because the high share of wind power serves to legitimize electricity use.

---

[3] https://efkm.dk/ministeriet/aftaler-og-politiske-udspil/energiaftalen/.

Equally, a missing focus on energy savings is reflected in the lack of measures to reduce energy use in existing buildings. New buildings are constructed to high energy standards, but there is significant potential for energy renovation of the older housing stock, which would be a relatively cheap way to reduce carbon emissions (Klimarådet 2017). The need for thermal improvement of the existing housing stock can partly be seen as a paradoxical result of the previous success with CHP, which ensured cheap heating for many years (recently, some plants have had to raise prices because the increasing electricity production from wind reduces the demand for electricity from CHP plants and thus increases the price of heating).

The relatively high share of wind power in the electricity system and growing prospects of electrification of heating and transport call for preparations of the energy system for more flexibility. Since 2010, smart grid solutions and flexible demand have attracted much interest, involving research and experiments. The smart grid concept concentrates on the electricity system, but it is increasingly acknowledged that a low carbon transition must involve the coevolution of several other systems. The discourse thus tends to change towards smart energy systems (Lunde et al. 2016).

## The Role of Households in the Low Carbon Transition

For a long time, households have had access to various subsidies for retrofits, installation of solar panels and replacement of oil burners with heat pumps. Some subsidies are still available, for instance, mediated through the electricity distribution companies,[4] but the present level is low. The support for solar panels has been characterized by stop-go policies, as the impacts and expenses have been difficult to predict. The Energy Agency runs a webpage with advice on energy savings, renovation, subsidies, etc.,[5] but in practice, it does not appear to be a high priority for the state to involve households actively in the transition. Households are mostly seen as passive consumers that can be motivated

---

[4] https://efkm.dk/aktuelt/nyheder/2016/dec/ny-energispareaftale-paa-plads/.
[5] https://sparenergi.dk/.

through prices, including reduced tariffs on electricity and electric cars. Based on smart grid solutions, households are also expected to react to flexible electricity prices.

Considering the scale of the challenge ahead, it might be beneficial to involve households more actively in the energy system. Some citizens take on this task themselves and organize both energy savings and provision of renewable energy, however some initiatives may raise complex dilemmas. For instance, it is not always desirable from a systems perspective when households increase self-sufficiency (e.g. through solar panels, domestic wind turbines and batteries). Nevertheless, engagement is decisive for the promotion of overall change, and active municipalities may facilitate a reasonable fit between considerations of systems and engagement.

## Sustainable Energy Consumption Initiatives

It is evident from the above description that the Danish energy system, Danish national energy policies, and initiatives for energy reductions have developed and changed over time in such a way that several aspects have become interlaced. Dependence on rational choice mechanisms across policy, initiatives and evolving system configuration has resulted in complex, and sometimes contradicting dynamics (such as a push for cheap district heating and resulting lack of thermal insulation measures). It is therefore no surprise that energy demand dynamics are multidimensional and difficult to address. Equally, it is of no surprise that several, sometimes contrasting initiatives for energy reduction come about, as also highlighted in the previous section.

National sustainable energy consumption initiatives tend to either focus on technological optimization and consumer adoption of more energy efficient products (e.g. energy saving contests; various Danish Energy Agency campaigns), or an altogether different approach of establishing small eco-communities that develop their own independent grids and alternative lifestyles (e.g. eco-villages; transition towns). As mentioned earlier, municipalities hold the potential for establishing initiatives that enable broader, systemic changes as well as local engagement, and a few sustainable energy consumption initiatives have been developed with this consideration in mind. An example of one such initiative is given below.

## MY CLIMATE PLAN MIDDELFART:
## A GOOD PRACTICE EXAMPLE

As part of a broader climate commitment strategy, Middelfart Municipality initiated a somewhat alternative approach to energy renovation of private homes in 2011, called 'My Climate Plan'. The municipality had good experiences with energy renovations of public and commercial buildings through the so-called ESCO models. Here, Energy Service COmpanies help the owner of a building to invest in optimizing energy consuming installations within the building, paid by the obtained energy savings. This setup was, however, not commercially viable for small private households, which the municipality also wished to address in their climate policy. As a result, Middelfart Municipality developed the idea of an 'ESCO light' initiative, where the professional Energy Service COmpanies were substituted with a local strategic partnership, consisting of a number of local stakeholders (craftsmen, a bank, and utility companies, among others) in order to provide better incentives for homeowners in the municipality to carry out energy renovations of their homes (Westergaard 2011).

A key challenge was that municipalities have no persuasive or regulative planning instruments to mobilize energy renovations among private homeowners. As a result, the lack of energy optimization for small privately owned buildings represents a hindrance for reaching ambitious climate targets in many Danish municipalities. Numbers from Teknologirådet (2008) show that while 145,000 households renovated their kitchen, 165,000 renovated bathrooms, and many fitted new floors, put in new windows and renovated roofs, only 20,000 carried out energy retrofits. Formerly available national subsidies for energy renovations have almost disappeared, so incentives are low. The strategic lever developed in Middelfart was built on the foundation that energy retrofits can be meaningful for homeowners to perform if investment in energy optimization is carried out together with general renovation projects in the house. For example, it is much cheaper to insulate a roof, if you are already installing a new one. The municipality saw the potential of using local craftsmen as spokespersons for energy renovations, when bidding on general renovation projects. Together with the Knowledge Center for Energy Savings in Buildings, the municipality helped to provide a supplement to an existing energy advice training that was offered to local craftsmen on commercial terms. The municipality facilitated meetings and dialogue with the local craftsmen in order to inspire them to take the training at the Knowledge Center, and provided a framework on how to approach private homeowners with energy saving advice. The course introduced the local craftsmen to energy related aspects of building renovations across different professions. Based on this course, the craftsmen advised homeowners about when it would be suitable to think about energy optimizations, and they started recommending each other, when they saw that complementary services were needed. The project gained political endorsement in the municipality due to the perspective of strengthening the business plans of local craftsmen.

Besides the partnership with the local craftsmen and the Knowledge Center, the municipality also developed a framework concerning further financial incentives for the homeowners. This involved the local utility company that could reimburse the private homeowners with 1 DKK per saved kWh, if the local craftsmen filled out a quality documentation of the energy savings (Escommuner 2013; Westergaard 2011). This reimbursement was based on the commitment that utilities have towards the Danish state to apply a certain amount of their income to carry out energy saving initiatives. A local bank agreed to provide more attractive loans for private homeowners that needed finance to carry out this kind of energy renovation (Jensen and Quitzau 2017). This was driven by the prospect of increased local opportunities and a possibility to market their local embeddedness and sustainability profile through involvement in the project.

By shifting the focus towards multi-actor engagement including craft professions, financial services and utilities, the municipality has developed a more 'structural' perspective that acknowledges renovation practices among homeowners. The case represents a good example of how such perspectives can include and address several complex situations related to energy renovations of private homes.

## CONCLUSION

Although the Danish electricity system has come to utilize high levels of wind power over the years, and the Danish heating system to a large extent is based on district heating, Denmark still needs considerable efforts to meet the targets of the Paris Agreement (Klimarådet 2018). Much of the national carbon-reduction measures focus on energy-intensive industrial plants (covered by the EU Emissions Trading System, ETS), through the establishment of offshore wind farms and other investments in wind, solar power and biogas.[6] Little is included to reduce emissions from the sectors that are not covered by ETS, such as transport, agriculture and buildings. Equally, a missing focus on energy savings is reflected in the lack of measures to reduce energy use in existing buildings. New buildings are constructed to high energy standards, but there is significant potential for energy renovation of the older housing stock, which would be a relatively cheap way to reduce carbon emissions. Finally, households have had access to various subsidies for retrofits, installation of solar panels and replacement of oil burners with heat pumps, however, although some subsidies are still available (often mediated through the electricity distribution companies[7]), the present level is low. The support for solar panels has been characterized by stop-go policies, as the impacts and expenses have been difficult to predict. The Energy Agency runs a webpage with advice on energy savings, renovation, subsidies, etc.,[8] but in practice, it does not appear to be a high priority for the state to involve households actively in the transition. The focus on energy reductions, particularly related to household energy demand is ultimately low, and the energy transition potential is seen mainly as a technical problem with technical solutions. Very little is done to target levels of energy demand at a national level. The presented good-practice example is a rare, but good, example of how everyday life perspectives, matters of convenience and institutionalized conditions around energy retrofitting potentials (or lack thereof) have been addressed at a municipal level.

[6] https://efkm.dk/ministeriet/aftaler-og-politiske-udspil/energiaftalen/.
[7] https://efkm.dk/aktuelt/nyheder/2016/dec/ny-energispareaftale-paa-plads/.
[8] https://sparenergi.dk/.

# REFERENCES

Christensen, T. H., Godskesen, M., Gram-Hanssen, K., Quitzau, M.-B., & Røpke, I. (2007). Greening the Danes? Experience with consumption and environment policies. *Journal of Consumer Policy, 30,* 91–116.

Dansk Fjernvarme. (2018). http://www.danskfjernvarme.dk/viden-om/fjernvarmeinfo.

Dietrich, O. W., & Morthorst, P. E. (2016). Danmark—Energiforsyning. In *Den Store Danske.* Gyldendal. http://denstoredanske.dk/index.php?sideId=70753.

Energistatistik. (2017). https://ens.dk/sites/ens.dk/files/Statistik/pub2017dk.pdf.

Energistyrelsen. (2018). https://ens.dk/ansvarsomraader/varme/information-om-varme.

Escommuner. (2013). https://escommuner.middelfart.dk/.

Hvelplund, F. (2007). Fra fælleseje til fjernejerskab og monopolkontrol. In E. Christensen & P. Christensen (Eds.), *Fælleder i forandring.* Aalborg Universitetsforlag.

Jensen, C. L., & Quitzau, M. (2017). Towards more eclectic understandings of energy demand and change—A tale of sense-making in the messiness of transformative planning. *Energy Research and Social Science, 31,* 253–262.

Klimarådet. (2017). *Omstilling frem mod 2030. Byggeklodser til et samfund med lavere drivhusgasudledninger.* https://www.klimaraadet.dk/da/rapporter/omstilling-frem-mod-2030.

Klimarådet. (2018). *Biomassens betydning for grøn omstilling.* https://www.klimaraadet.dk/sites/default/files/downloads/klimaraadet_biomassens_rapportno4_digi_01.pdf.

Lunde, M., Røpke, I., & Heiskanen, E. (2016). Smart grid: Hope or hype? *Energy Efficiency, 9,* 545–562.

Teknologirådet. (2008). *Klimarigtigt byggeri – vi kan, hvis vi vil!* [Climate-friendly buildings – We can do it, if we want to!]. http://www.tekno.dk/wp-content/uploads/2014/12/p08_miljoebyg_rapport.pdf. (Old name: Teknologirådet; New name: Fonden Teknologirådet).

Westergaard, M. (2011, February). *ESCO-Light i Middelfart Kommune* [ESCO-Light in Middelfart Municipality] (pp. 32–33). Silkeborg: Teknik & Miljø.

Wistoft, B., Thorndahl, J., & Petersen, F. (1992). *Elektricitetens Aarhundrede. Dansk elforsynings historie. Bd. 2. 1940–1991.* Danske Elværkers Forening.

# CHAPTER 5

# Reducing Residential Carbon Emissions in Ireland: Challenges and Policy Responses

*Gary Goggins, Frances Fahy and Eimear Heaslip*

**Abstract** Carbon emissions from the residential sector in Ireland are higher than the European average and are rising. This is a concern in a country already struggling to meet its agreed climate targets. In this chapter, the authors highlight key trends that underpin household energy use in Ireland and undertake a critical examination of related energy policy, with particular attention to the role of the consumer. They find a broad objective to place the consumer at the forefront of Ireland's energy transition, but specific detail of how complex social and technical changes will be realised are lacking. The chapter concludes with a case study demonstrating how sustainable energy initiatives can bring together multiple actors with the common aim to address fuel poverty and lower carbon emissions.

G. Goggins (✉) · F. Fahy · E. Heaslip
School of Geography and Archaeology and Ryan Institute,
National University of Ireland Galway, Galway, Ireland
e-mail: gary.goggins@nuigalway.ie

F. Fahy
e-mail: frances.fahy@nuigalway.ie

E. Heaslip
e-mail: eimear.heaslip@nuigalway.ie

© The Author(s) 2019
F. Fahy et al. (eds.), *Energy Demand Challenges in Europe*,
https://doi.org/10.1007/978-3-030-20339-9_5

47

**Keywords** Energy policy · Sustainable consumption · Residential carbon emissions · Ireland · Transition

## INTRODUCTION

Ireland's energy system is heavily reliant on fossil fuels and largely dependent on imports, primarily of oil and gas. The contribution of renewables in the energy mix is significantly behind Ireland's agreed target under EU Directives. Additionally, Ireland has been identified as a European 'laggard' in reducing non-Emission Trading Scheme sector emissions (i.e. emissions associated with energy use in buildings and in transport, and emissions from agriculture). On a macro level, Ireland will need to introduce substantial changes in order to reach emissions targets for 2030 and beyond. This responsibility must be shared across different sectors in society, including the residential sector, which currently accounts for a quarter of all emissions. Indeed, Irish homes on average use more energy and emit substantially more $CO_2$ than their European counterparts (SEAI 2018). There are significant opportunities for reducing household energy use through technological advances, social innovation and changing user practices.

## SOCIO-MATERIAL DYNAMICS OF HOUSEHOLD ENERGY USE IN IRELAND

The residential sector is responsible for 25% of energy use in Ireland, second only to transport, at 42% (SEAI 2018). In 2015, the average household in Ireland used 7% more energy than the EU average and emitted almost 60% more $CO_2$ than the average EU home (SEAI 2018). Continuing dependency on high-carbon fuels (e.g. oil, coal, peat), falling oil prices and higher incomes are some of the factors that contribute to recent increases in $CO_2$ emissions across the residential sector. In addition, Ireland has experienced a growth in population of 25% over the period 2000–2016, with upward trends forecast to continue. To accommodate an expanding population, the number of dwellings has also increased to currently stand at 1.7 million households, although this figure remains significantly short of projected housing need, with annual average demand estimated to be up to 36,000 units for the next 30 years (IBEC 2018). Yet despite the number of new homes built to ever increasing energy performance standards, the Irish housing stock is among the poorest in Europe in terms of energy efficiency (Goggins et al. 2016). Current trends also show that households are getting bigger, with an average 15% increase in floor area across all homes between 2000 and 2016 (SEAI 2018). Another contributing factor in residential energy use is the number of persons per household, which for Ireland is the second highest in Europe at 2.7 persons per dwelling (Eurostat 2018).

The location of dwellings can have a significant impact on the type of fuel and heating systems available to households, as some options such as connection to a mains gas network or district-heating system may not be feasible in rural areas. The use of heating oil is particularly dominant in rural households, and the carbon-intensive practice of harvesting and burning peat is socially and culturally engrained in many rural locations. Housing type is also a significant material factor in affecting residential energy use, particularly regarding space heating due to the level of exposed surface area, which is generally greatest for detached houses and lowest for apartments. In Ireland, detached houses make up the vast majority of dwellings in rural areas (83%), compared to 19% in urban locations and 42% of the overall housing stock. Just 7% of people in Ireland live in apartments or flats, which is easily the lowest proportion across the EU (Eurostat 2015). While there are clear opportunities for reducing energy use through increasing thermal efficiency of homes, other socially oriented policy instruments such as those that might reduce per capita living area (e.g. encouraging downsizing after children move out) or sharing of resources (e.g. common laundry rooms) can also reduce energy use, but are less developed in Ireland.

## Energy Policy in Ireland

Ireland's energy policy is largely techno-centric, with a strong emphasis on technological change and innovation. For example, improving energy efficiency in the residential sector is considered a critical element of Ireland's energy policy and of a sustainable energy transition, with the Sustainable Energy Authority of Ireland (SEAI) estimating a capital investment of the order of €35 billion over 35 years would be required to make the existing housing stock low carbon by 2050 (SEAI 2017a). Traditionally, Ireland's energy policy was considered the domain of (centralised) Government and utility companies, with other actors (e.g. local authorities; business) largely sidelined, and consumer participation assuming a passive role. However, recent shifts in the framing of the energy challenge have positioned consumers at the forefront of Ireland's energy transition (Mullally et al. 2018).

The most recent government long-term energy policy *White Paper* sets out a vision to 2030 where Ireland's energy system 'will become more decentralised, altering many traditional assumptions about demand and supply', requiring 'deep change in the mindsets of individual consumers, businesses, agencies, and utility companies' (DCENR 2015: 5). This outlook aligns with the prevailing narrative in European energy policy that a transition to a low-carbon society requires integrated action from a broad range of actors, including householders (Genus et al. 2018). The central role of householders is indirectly elaborated in the *White Paper*, insofar as it proposes that citizens move from being 'passive consumers' to 'active citizens', and that every citizen has a role to play in the energy transition. Although the paper neglects to define what is understood by 'active citizen', it does state that consumer choice—in the home, in the community, at work and when travelling—is perceived as an important aspect of the energy citizen's role and responsibilities (DCENR 2015). Such a perspective indicates an individualised approach to how energy demand is problematised, with an implied duty upon consumers to make the 'better' energy choices (Genus et al. 2018). While this perspective might evoke traditional problem framings prevalent in EU energy policy around energy efficiency, rational choice and behaviour change, the paper also proposes that 'landowners, neighbours and communities will be able to engage with infrastructure providers and local government to ensure acceptable outcomes for all energy users' and become more engaged in the energy landscape in Ireland (DCENR 2015: 40). Although it is not made explicitly clear how this inherently complex social and technical change might come about, this collective approach acknowledges interactions between various actors as key to a sustainable energy transition, thereby presenting opportunities for achieving long-term sustainability goals through changes in complex interactions (Jensen et al. 2018).

## TRENDS IN NATIONAL HOUSEHOLD ENERGY CAMPAIGNS IN IRELAND

Prominent national energy campaigns are reflective of Ireland's energy policy and associated underlying notions of how, and what kind of, change might come about. Campaigns generally focus on two main areas, encouraging retrofitting of homes and increasing energy awareness, both of which fall under popular problem framings in EU energy policy that prioritise energy efficiency through system optimization, consumer choice and behaviour change (Jensen et al. 2018). For example, the SEAI have run a 'Tips and Advice Campaign' and a 'Be your own energy manager' campaign, focused on providing householders with a series of steps on how to reduce their energy consumption. Recommendations centre around technical innovation and optimization, including using timers with hot water and heating systems, and behavioural changes such as ensuring heating and hot water systems are only switched on as required. Retrofitting is also encouraged, and information campaigns are generally aimed at encouraging householders to avail of significant grants for energy efficiency and renewable energy upgrades. Over recent years, grant aid for households to engage in energy efficiency improvements (e.g. cavity insulation; solar photovoltaic systems) is offered through one of several schemes run by the SEAI. Applications can be made by individual households or as part of a community scheme, with various funding rates available depending on a number of predefined socio-economic and other criteria such as the age of dwelling. The SEAI has also undertaken research into human and psychological factors that influence uptake of such schemes, with a particular focus on encouraging householders to retrofit and the barriers thereof (SEAI 2017b). In total, over 375,000 homes received government grants for energy efficiency improvements between 2000 and 2016.

A review of recent Sustainable Energy Consumption Initiatives (SECIs) in Ireland according to their problem framing reflects the

**Table 5.1**   Number of national SECIs according to their problem framing

| Problem framing | No. of initiatives |
|---|---|
| Changes in complex interactions | 9 |
| Changes in everyday life situations | 6 |
| Changes in individuals' behaviour | 15 |
| Changes in technology | 25 |

dominance of traditional problem framings that prioritize changes in technology, and changes in individuals' behaviour (Table 5.1). At the same time, there is ample evidence of initiatives that understand the challenge of changing energy use as a more complex and collective concern. These initiatives seek to bring about long-term systemic change by targeting changes in everyday life situations or changes in complex interactions. One such example is the SHARE project, a partnership between sustainable energy organisations working with social housing providers and residents in eight European regions in the UK, Bulgaria, Estonia, France, Germany, Ireland, Slovenia and Sweden.

## Case Study: SHARE (Social Housing Action to Reduce Energy Consumption)

Within an overall context of reducing carbon emissions and reducing the risk of fuel poverty, the SHARE project (2006–2008) aimed to increase awareness of the opportunities and practical options for sustainable energy retrofit and behavioural change. Focusing on existing housing, the project targeted low-income groups located in 10 distinct communities of different scales and sizes. The project was coordinated by the Severn Wye energy agency in the UK, and included eight partners including the Irish-based Tipperary Energy Agency (TEA).

The TEA engaged with a range of stakeholders in the public and voluntary sector, including householders living in social housing, designers and implementers of social housing, and local authorities that are responsible for the management of social housing in Ireland. The project used several dissemination techniques including training sessions, information provided on the project website, telephone interactions, information stands at Local Authority events and site visits.

### Methods for Intervention

SHARE Forums were set up for each of the eight countries involved to promote good practices and encourage the sharing of experiences. Forums included social housing providers, residents, local authorities, energy providers, building and services contractors, and a variety of specialists working within the sector. Training sessions were undertaken with both the householders and those that are responsible for managing, designing and building social housing. Awareness and advice plans on existing materials and good practices for each participating country were produced and a series of case studies covering the Forums training and awareness campaigns were made available on the project website.

The TEA were responsible for the training sessions in Ireland and concentrated on providing information related to insulation, more energy efficient central heating systems, optimised heating controls, smart metering equipment and renewable technologies where appropriate. Householders were required to attend the training workshops and then attempt to implement some of the recommendations to reduce their energy use and lower energy costs. To assist householders, the project involved face-to-face visits at participants' homes and provided needs-based tailored information and supports (Heiskanen et al. 2018). The SHARE Forums identified some key areas that tenants typically have problems with including understanding energy bills, efficient use of heating and hot water controls, and awareness and management of energy related to electrical appliances and lighting. Other key areas particularly relevant to Ireland included ventilation and condensation, draught proofing, insulation, fuel poverty, options for home heating, renewable energy options, and grants and assistance.

### *Framing the Energy Challenge*

As the project focused on low-income households, the TEA primarily focused on framing energy in financial terms, for example 'save on your energy bills'. Broader environmental implications of saving energy were also discussed with participants. Topics of comfort or health related to energy use were not addressed, however these were subsequently recognised as important factors that might be targets for future projects. From the participants' perspective, energy was framed in terms of expensive energy bills, and discussion about the lack of support from government would arise from time to time in the Forums. The implementers reported that trying to focus specifically on energy in social housing presents a challenge, as people have lots of other issues in their lives that they need to deal with, and energy is not often a priority. Hence, one of the challenges was that householders would tend to voice other issues affecting them in the community or housing estate, as they had no other platform in which to voice their concerns.

### *SHARE Project Impact*

While there were no studies conducted of the environmental or monetary impact of the project, 85% of participants reported the training to be 'very useful', 15% found it to be 'fairly useful', with no participant indicating that the training was 'of no use'. The average feedback rating given by 89 participants was 4.5 (out of 5). As there was no consumption data recorded, it is difficult to determine whether there were significant, or any, changes in energy use in the long term. However, the project was successful in bringing together a range of actors that influence the consumption of energy in social housing, and promoting collaboration and knowledge exchange between these stakeholders. Although the energy challenge was framed in financial terms, the initiators understood energy use as an outcome of complex interactions between different activities, professions and sectors. The project involved several methods of intervention, and promoted the sharing of responsibility between multiple actors including householders, social housing project designers, implementers and government.

## Conclusion

Ireland has a long way to go in order to meet its binding carbon emissions targets for 2020 and beyond. If Ireland is to meet its longer-term carbon emissions targets, household energy use and related carbon emissions will need to reduce dramatically. This chapter provides examples of recent SECIs undertaken in Ireland. As evidenced in Table 5.1, the majority of these initiatives target changes in individual behaviour or technological changes. These approaches mirror general government policy, where technological and innovation approaches are pursued to provide 'solutions' for problems such as excessive energy use. However, recent trends suggest that technological approaches alone are insufficient to deliver the necessary reductions in residential carbon emissions.

The good practice example illustrates some of the complexities in achieving more sustainable energy use. The SHARE initiative brought together a range of actors from across society, including local authorities, householders and practitioners, to help people in social housing to reduce their energy use and alleviate energy poverty. The tailored approach aimed to overcome some of the contextual difficulties experienced by householders and other actors. However, it also demonstrated the myriad other difficulties facing low-income households, and showed that people are somewhat detached from their energy use. The project also identified some shortcomings in existing SECIs, such as linking energy use with related issues such as comfort, health and well-being, as well as environmental and economic concerns. This more holistic approach to energy use should be considered in designing future initiatives.

## References

DCENR. (2015). *Ireland's transition to a low carbon energy future 2015–2030.* Available at https://www.dccae.gov.ie/documents/Energy%20White%20Paper%20-%20Dec%202015.pdf. Accessed 27 April 2018.

Eurostat. (2015). *EU Statistics on Income and Living Conditions (EU-SILC).* Available at http://ec.europa.eu/eurostat/statistics-explained/index.php/Housing_statistics. Accessed 20 November 2018.

Eurostat. (2018). *Eurostat labour force survey.* Available at https://ec.europa.eu/eurostat/statistics-explained/index.php/Household_composition statistics#Household_size. Accessed 20 November 2018.

Genus, A., Fahy, F., Goggins, G., Iskandarova, M., & Laakso, S. (2018). Imaginaries and practices: Learning from 'ENERGISE' about the integration of social sciences with the EU Energy Union. In *Advancing Energy Policy* (pp. 131–144). Cham: Palgrave.

Goggins, J., Moran, P., Armstrong, A., & Hajdukiewicz, M. (2016). Lifecycle environmental and economic performance of nearly zero energy buildings (NZEB) in Ireland. *Energy and Buildings, 116,* 622–637.

Heiskanen, E., Laakso, S., Matschoss, K., Backhaus, J., Goggins, G., & Vadovics, E. (2018). Designing real-world laboratories for the reduction of residential energy use: Articulating theories of change. *GAIA-Ecological Perspectives for Science and Society, 27*(1), 60–67.

IBEC. (2018). *Better housing: Improving affordability and supply.*

Jensen, C. L., Goggins, G., Fahy, F., Grealis, E., Vadovics, E., Genus, A., et al. (2018). Towards a practice-theoretical classification of sustainable energy consumption initiatives: Insights from social scientific energy research in 30 European countries. *Energy Research & Social Science, 45,* 297–306.

Mullally, G., Dunphy, N., & O'Connor, P. (2018). Participative environmental policy integration in the Irish energy sector. *Environmental Science & Policy, 83,* 71–78.

SEAI. (2017a). *Ireland's energy projections: Progress to targets, challenges and impacts.* Available at https://www.seai.ie/resources/publications/Irelands_Energy_Projections.pdf. Accessed 18 April 2018.

SEAI. (2017b). *Behavioural insights on energy efficiency in the residential sector.* Sustainable Energy Authority of Ireland.

SEAI. (2018). *Energy in the residential sector (2018 report).* Available at https://www.seai.ie/resources/publications/Energy-in-the-Residential-Sector-2018-Final.pdf. Accessed 2 May 2018.

# CHAPTER 6

# UK Responses to the Energy Challenge: Dominant Framings and New Imaginaries

*Marfuga Iskandarova and Audley Genus*

**Abstract** The chapter identifies dominant framings of contemporary energy challenges and possible responses in relation to developments in the UK. It summarises national trends in energy consumption and the material, societal and policy factors that influence them. Examples are provided of energy campaigns and sustainable energy consumption initiatives that illustrate different problem framings of energy challenges. A 'good practice' example of a UK initiative that involves changes in complex interactions demonstrates the value of the complex approach that targets energy use along with other aspects of sustainable living. The chapter concludes by pointing to an alternative framing and imaginary that could tackle climate change more effectively.

**Keywords** 'BedZED' · Energy policy · Good practice · Imaginaries · Problem framings

M. Iskandarova (✉) · A. Genus
Kingston University, London, UK
e-mail: m.iskandarova@kingston.ac.uk

A. Genus
e-mail: A.Genus@kingston.ac.uk

© The Author(s) 2019                                                          59
F. Fahy et al. (eds.), *Energy Demand Challenges in Europe*,
https://doi.org/10.1007/978-3-030-20339-9_6

## Introduction

This chapter outlines developments in UK national energy policies and Sustainable Energy Consumption Initiatives (SECIs), exploring dominant framings for current energy challenges and possible solutions. The chapter identifies approaches that rely on alternative problem framings (or 'imaginaries' c.f. Genus and Iskandarova 2018; Jasanoff and Kim 2009) that may be better adapted to addressing these challenges.

The chapter consists of six sections. The section "Socio-Material Dynamics of Domestic Energy Use" introduces socio-material aspects of energy use in households. The next section, "Energy Policy in the UK", provides a brief summary of nationally specific trends and their manifestation in energy policy in the UK with implication for energy consumption. The section "Trends in Energy Campaigns and Types of Sustainable Energy Consumption Initiatives in the UK" gives examples of energy campaigns and SECIs that illustrate the different problem framings of energy discussed in Chapter 2 (see also Jensen et al. 2017). The section "BedZED: A 'Good Practice' Example" provides a *good practice* example of a national SECI that corresponds to the 'changes in complex interactions' category in the problem framing typology. Finally, the section "Concluding Remarks" summarises how the energy challenge is currently typically framed in the UK, while pointing to an alternative imaginary or framing which could tackle the challenge more effectively.

## Socio-Material Dynamics of Domestic Energy Use

According to the Department for Business, Energy & Industrial Strategy, the domestic sector accounts for more than a quarter of total energy consumption in the UK (28% in 2017). The fuel mix has changed significantly since 1970, when 49% of final energy consumption was provided by solid fuels and 24% by gas; these days the balance looks very different— 1.6 and 66% respectively (BEIS 2018). In the UK, space and water heating accounts for 80% of final domestic energy consumption, which is also susceptive to temperature fluctuations. In addition to weather-related factors, energy consumption is affected by household characteristics including the age of housing, efficiency measures (e.g. level of insulation) and the usage of appliances (BEIS 2018). Air conditioning is not a common feature in British houses; fireplaces as well as outdoor heaters and power showers are more popular. An additional factor to be considered is the level of comfort required, i.e. a reasonable level of warmth, which varies over time. The average room temperature in the UK is around 18 degrees (compared with 12 degrees in 1970), and, on average, UK homes are heated for 8 hours per day in winter. About 70% of UK homes with central heating heat their homes twice per day and occasionally boost the heating when required (OVO Energy 2018). Energy prices and relative incomes also impact consumption levels. In the period since 2005, gas and electricity prices have more than doubled. This significant increase in fuel prices, combined with an economic downturn, is likely to reduce consumption as consumers became more conscious of their household budgets (BEIS 2018).

The UK housing stock is old compared with most European countries. As a result, many houses have poor insulation, resulting in additional consumption to maintain a given level of comfort. Nonetheless, older housing stock is being gradually replaced with newer, more energy efficient homes. Therefore, there have been some key changes to household characteristics, as well as energy efficiency measures (e.g. more efficient boilers; installation of double glazing; cavity wall insulation), which have put downward pressure on energy use (BEIS 2018).

People in the UK prefer living in houses to flats. This is partially explained by assumptions about respect for privacy and independence, and pride in ownership. Additionally, houses are typically 'freehold', i.e. includes the ownership of both the building and the land, in contrast with flats or apartments, which in England and Wales are most commonly owned on a leasehold basis. The UK is also the only EU country not to have minimum-space standards for homes; as a result, it has the smallest new homes in Europe, significantly smaller than 100 years ago (Henley 2012).

## Energy Policy in the UK[1]

UK energy policy today seeks to deliver solutions to the so-called energy 'trilemma'—the perceived need for secure, affordable and clean energy supplies for the UK's economic success. Although the UK has pursued a centralised approach to energy for many decades, the government and other interest groups intend to develop decentralised energy and storage systems

---

[1] This chapter was written at the time when the UK was in the process of Brexit negotiations. There may be changes to energy and climate areas when the UK leaves the EU.

and replace significant volumes of large, transmission-connected fossil-fuel power stations with smaller, often distribution-network-connected, renewable generation technologies such as wind and solar. This fundamental shift will have implications for how the energy system is operated.

A catalyst for the growth of renewables is the legal requirement that the UK provide at least 15% of its energy from renewable energy sources by 2020, with the Department of Energy and Climate Change (DECC) being established in 2008 to deliver this target. The Climate Change Act 2008 is part of the UK government's plan to reduce greenhouse gas emissions. The UK Renewable Energy strategy 2009 was instituted as an action plan for delivering the UK's renewable energy objectives, and the Feed in Tariff (FiT) scheme was launched in 2010 as a policy mechanism to accelerate investment in renewable energy.

Through its Microgeneration Strategy, launched in 2011, and the Renewable Heat Incentive, the coalition government put in place a range of financial incentives to encourage the deployment of small-scale, on-site, renewable energy. This was followed by the announcement of the Green Deal Scheme, a programme for building refurbishment i.e. energy-saving improvements (the government scheme was closed in 2015 and relaunched in 2017 as the Green Deal Finance company backed by private investors). The energy efficiency agenda was underpinned by the Energy Efficiency Strategy of 2012, which set the direction for energy efficiency policy, and identified steps to stimulate the energy efficiency market.

The UK government's first ever Community Energy Strategy was launched in 2014. It aimed to encourage communities to play a greater role in achieving energy and climate change goals, e.g. community involvement in generating electricity. Recent years have seen a growth in small-scale installations of renewable energy aided by the UK FiT, but since the UK general election in 2015, there have been substantial, negative changes to support for key renewable energy technologies (e.g. significant reduction of FiT).

The electricity market reforms aimed to attract investment needed to replace and upgrade the UK's electricity infrastructure and enable it to meet the growing demand for electricity. One of the key mechanisms of the reform is Contracts for Difference. It is designed to support investment in new low-carbon generation, with a technology-dependent fixed price (BEIS 2015/2017). The reform was underpinned by the Energy Act 2013, which aimed to maintain a stable electricity supply as coal-fired power stations are retired.

A recent development aimed to promote the deployment of innovative technologies in the design of an electricity system based on smart metering and supporting infrastructure. The government is committed to ensuring that smart meters are offered to every home and small business by the end of 2020, enabling smart tariffs and other benefits for consumers, arguably putting consumers in control of their energy use.

Given that fuel poverty affects over 4 million UK households (roughly 15% of all households), it is not surprising that fuel poverty and energy efficiency are the focal points of policy discourse and energy campaigns in the UK. There have been a number of government schemes in recent years aimed at reducing fuel poverty: the Warm Front Scheme ran until January 2013, its replacement the Energy Company Obligation (ECO) scheme (and a subsequent version known as the Affordable Warmth Obligation) began in early 2013; it offers grants for energy-saving improvements to people's homes, such as insulation works or heating system upgrades. The Social Fund Cold Weather Payments scheme runs from 1 November to 31 March each year for those receiving certain benefits payments in Great Britain (e.g. Pension Credit, Income Support, Universal Credit, Support for Mortgage Interest). The Warm Home Discount Scheme provides discounts on electricity or gas bills during the winter months for those receiving Pension Credit and those on a lower income. A Winter Fuel Payment also helps those who qualify for the scheme pay heating bills. The 2015 Fuel Poverty Strategy for England aims to improve the homes of the fuel poor by 2030 achieving where possible a minimum energy efficiency rating of Band C.

## TRENDS IN ENERGY CAMPAIGNS AND TYPES OF SUSTAINABLE ENERGY CONSUMPTION INITIATIVES IN THE UK

Concerns about fuel poverty and energy efficiency are at the heart of energy campaigns and some initiatives in the UK. They often target energy users by providing information and advice regarding energy bills and choosing a supplier, particularly for low-income households. Between 2012 and 2015, the Energy Bill Revolution/Campaign for Warm Homes & Lower Bills movement aimed to raise public awareness about the UK's cold home crisis, and to gain support for making home energy efficiency an infrastructure investment priority that would also help end fuel poverty, reduce carbon emissions and create green jobs. Other campaigns, such as the Clean British Energy campaign (run by Friends of the Earth) are more concerned with moving away from reliance on fossil fuels and cutting carbon from the energy system. The campaigns sponsored by energy companies such as British Gas, EDF Energy, E.ON, npower, ScottishPower and SSE aim to cut the number of deaths and injuries caused by carbon monoxide poisoning.

The most prevalent type of SECIs in the UK are those directed at changing individual behaviour or involving energy efficiency-related technical change—a tendency that accords with findings across the countries investigated as part of the ENERGISE project. Energy efficiency and reduction of energy use and carbon emissions (resulting in carbon-neutral or low-carbon living) are declared as the main objectives pursued by SECIs. The issue of fuel poverty is still addressed directly only by a handful of the SECIs. This can be partly explained by the fact that SECIs are usually carried out by communities with sufficient resources for investing in those initiatives. Smart metering and use of technology for monitoring energy consumption and emissions are among priorities for the UK SECIs. A community engagement element and an inclusive approach are important for many UK energy initiatives, e.g. in community renewable energy projects, which often represent active involvement of citizens who participate in local electricity generation.

Although citizens in the UK are often portrayed as passive energy consumers for whom policy-makers attempt to deliver 'affordable' energy and competitive markets, the overview of SECIs in the UK illustrates the potential to address issues of domestic energy consumption somewhat differently. The example presented in the following section demonstrates how the energy challenge is framed and addressed as an essential part of a contemporary sustainable living concept.

## BedZED: A 'Good Practice' Example

The good practice example discussed here, the 'BedZED' development in a suburb of London, is among initiatives that aim to change 'complex interactions' in relation to energy. BedZED is a short name for Beddington Zero (Fossil) Energy Development. There are several such eco-developments in the UK, developed by Bioregional Development Group, which was founded as a registered charity in 1994 by environmental activists concerned about the effects of unsustainable consumption on the environment.

BedZED, a purpose-built eco-village, is located in the borough of Sutton, in south London. The developer's website describes it as the 'UK's first large-scale, mixed use sustainable community with 100 homes, office space, a college and community facilities'. Completed in 2002, it has dwellings of various sizes and tenures, which include 82 houses, 17 apartments and 1405 m² of workspace. The aims of the development are stated to be: to show what a 'truly sustainable community looks like'; to reduce ecological footprint of contemporary living and reduce carbon emissions related to consumption of heating and lighting, water, food, transport and waste. In addition to the sustainability of the finished BedZED product, every aspect of construction was considered in terms of its environmental impact. Materials used in construction were selected for low environmental impact, sourced locally where possible and from reclaimed and recycled materials where feasible.

Criteria for measuring performance include monitoring electricity, heat and water consumption; car ownership and miles travelled, air miles travelled, bicycle ownership; number of households who grow their own food; organic vs non-organic food consumption; recycling rates; proportion of the foregoing in, and the total carbon footprint of BedZED. Though it is prudent to exercise caution in interpreting any ongoing, and as yet relatively short-lived phenomenon, the BedZED initiative can be considered to be a success. A survey in 2007 showed that BedZED's total energy consumption was 82.4 kWh/m²/year, compared with a UK residential total of 275.3 kWh/m²/year; monitoring showed that BedZED households used 2579 kWh of electricity per year, which is 45% lower than the average in Sutton (Hodge and Haltrecht 2009). BedZED (2007) related carbon emissions were 19.9 $CO_2$/m²/year, compared with the UK average of 63.3 $CO_2$/m²/year (based on dwellings built in 2002). Between 2012 and 2015, BedZED's annual gas consumption was 36% lower than a typical conventional development in Sutton of the same size and mix, and the annual electricity consumption was 27% less. An on-site car club, ample secure cycle parking, good public transport links and only 0.6 car parking spaces per home resulted in reduction in travel-related greenhouse gas emissions, which was 53% less $CO_2$ eq than the UK average (Schoon 2016). An earlier 2007 survey found that 17% of BedZED residents travelled to work by car, compared with the then Sutton average of 49%.

BedZED residents share an understanding of the initiative as a template for more sustainable living based on the unique design of the homes (e.g. allows using more of daylight, passive ventilation, passive solar gain), sense of community, the garden and conservatory-like 'sunspaces', the 'green' features of homes, the pioneering car club and reduced energy bills. Monitoring shows that sustainable lifestyles account for around half the eco-savings at BedZED, and energy demand is dramatically reduced compared with an equivalent conventional development. Small-scale, on-site energy generation included a biomass CHP plant (this hadn't reached the agreed outputs and was replaced in 2005 by three conventional natural gas-fired boilers) and photovoltaic panels (Schoon 2016). Homes are fitted with energy efficient and water-saving appliances, visible meters and 'super' insulation. Critics of the initiative would, however, point to the high cost of completing the development, and problems with the originally envisaged on-site energy plant and water treatment facilities. Although the energy consumption in the homes is much lower than average, residents living in BedZED are unable to get to a 'one planet living level'; residents had an average ecological footprint of about 2.5 planets' worth (with a potential to reduce it down to 1.9–1.7 planets) (Hodge and Haltrecht 2009).

Nonetheless, BedZED demonstrates the value of a complex approach that targets energy use along with other aspects of sustainable living (e.g. water use, transport, waste). Professional design (though not by citizens) combined with financial contribution from households provided a winning combination of expertise and involvement of residents. In the BedZED example, energy use is treated as an outcome of material and social organisation; an environment that is susceptible to more sustainable practices is created, and community building is seen as a crucial element of a sustainable living initiative. The BedZED example suggests that a more holistic approach to sustainable energy could be effective if adopted by policy-makers. Supporting sustainable/eco developments, where energy is addressed and understood in the context of related sustainable practices, could make sustainable living (including energy consumption) more attractive and easier to achieve.

## Concluding Remarks

It is recognised by the UK government that the participation of a diverse range of actors can facilitate the development of a more efficient smart and flexible energy system. In a recent consultation, BEIS announced the aim of the reform as maximising the ability of consumers to play an active role in managing their energy needs (BEIS/Ofgem 2016). However, the emphasis is on communicating effectively the benefits of smart meters and intelligent devices to manage energy use. This will not necessarily mean greater consumer engagement, and the focus of policy is still on reducing energy demand rather than citizens becoming 'prosumers'.

Overall, UK policy and other actors tend to frame the energy challenge and responses to it as requiring individual behaviour change or the diffusion of energy efficiency-related technology. The examples of UK initiatives featuring in the ENERGISE database support this viewpoint. However, there are cases that frame the challenge differently. For example, at BedZED unsustainable energy use is framed as a fundamental problem concerned with our way of life. Thus, energy use, practices and change are bound up with material and social organisation and the ways in which a community of people can live their everyday lives, rather than the outcome of individual actions or technology deployment.

## References

BEIS. (2015/2017). *Electricity market reform: Contracts for difference*, published 26 February 2015, updated 8 February 2017 [Online]. Available at: https://www.gov.uk/government/collections/electricity-market-reformcontracts-for-difference. Accessed 25 April 2018.

BEIS. (2018, July). *Energy consumption in the UK* [Online]. Available at: https://www.gov.uk/government/statistics/energy-consumption-in-the-uk. Accessed 10 January 2019.

BEIS/Ofgem. (2016, November). *A smart, flexible energy system. A call for evidence* [Online]. Available at: https://www.gov.uk/government/consultations/call-for-evidence-a-smart-flexible-energy-system. Accessed 25 April 2018.

Climate Change Act 2008. Available at: http://www.legislation.gov.uk/ukpga/2008/27/contents.

DECC. (2009). *The UK renewable energy strategy* [Online]. Available at: https://www.gov.uk/government/publications/the-uk-renewable-energy-strategy.

DECC. (2011, June 22). *Microgeneration strategy* [Online]. Available at: https://www.gov.uk/government/publications/microgeneration-strategy. Accessed 25 April 2018.

DECC. (2012, November 16). *Energy efficiency strategy* [Online]. Available at: https://www.gov.uk/government/collections/energy-efficiency-strategy. Accessed 25 April 2018.

DECC. (2014, January 27). *Community energy strategy* [Online]. Available at: https://www.gov.uk/government/publications/community-energy-strategy. Accessed 25 April 2018.

DECC. (2015, March 3). *Fuel poverty strategy for England* [Online]. Available at: https://www.gov.uk/government/speeches/fuel-poverty-strategy-for-england. Accessed 25 April 2018.

Energy Act 2013. Available at: http://www.legislation.gov.uk/ukpga/2013/32/contents/enacted. Accessed 25 April 2018.

Genus, A., & Iskandarova, M. (2018). *Policy paper 1: State of the art and future of policy integration for EU Policy on energy consumption.* ENERGISE—European Network for Research, Good Practice and Innovation for Sustainable Energy, Deliverable No. 6.4.

Henley, J. (2012, May 16). Why are houses in Britain so small? *The Guardian* [Online]. Available at: https://www.theguardian.com/global/shortcuts/2012/may/16/architecture-housing. Accessed 25 April 2018.

Hodge, J., & Haltrecht, J. (2009). *BedZED seven years on: The impact of the UK's best known eco-village and its residents.* Wallington: BioRegional. Available at: https://www.bioregional.com/resources/bedzed-7-years-on. Accessed 14 May 2019.

Jasanoff, S., & Kim, S.-H. (2009). Containing the atom: Sociotechnical imaginaries and nuclear power in the United States and South Korea. *Minerva, 47,* 119–146.

Jensen, C. L., Goggins, G., & Fahy, F. (2017). Construction *of typologies of sustainable energy consumption initiatives.* ENERGISE—European Network for Research, Good Practice and Innovation for Sustainable Energy, D2.4.

OVO Energy. (2018). *What's the average room temperature and thermostat setting in the UK?* [Online]. Available at: https://www.ovoenergy.com/guides/energy-guides/average-room-temperature.html. Accessed 25 April 2018.

Schoon, N. (2016). *The BedZED story: The UK's first large-scale, mixed-use eco-village.* Wallington: BioRegional. Available at: https://www.bioregional.com/resources/bedzed-the-story-of-a-pioneering-eco-village.

WEBSITES

BioRegional. BedZED. http://www.bioregional.com/bedzed/. Accessed 12 May 2017.

Energy Bill Revolution: The campaign for Warm Homes & Lower Bills. http://www.energybillrevolution.org/.

Energy UK. Media and Campaigns. https://www.energy-uk.org.uk/media-and-campaigns.html.

Green Deal: Energy saving for your home. https://www.gov.uk/green-deal-energy-saving-measures.

Help from your energy supplier: The Affordable Warmth Obligation. https://www.gov.uk/energy-company-obligation.

Social Fund Cold Weather Payments. https://www.gov.uk/government/collections/social-fund-cold-weather-payments#latest-release.

Warm Home Discount Scheme. https://www.gov.uk/the-warm-home-discount-scheme.

Winter Fuel Payment. https://www.gov.uk/winter-fuel-payment/eligibility.

CHAPTER 7

# Turning Off the Gas Tap: Sustainable Energy Policies, Practices and Prospects in the Netherlands

*Julia Backhaus*

**Abstract** This chapter describes efforts to transition to more sustainable ways of living in the Netherlands. Backhaus captures the status quo of Dutch sustainable energy policies and projects in clarity and brevity, suggesting that prospects to live up to the commitment made in the Paris Agreement are dim. The Perspective project, a major, yet not well-known Dutch research programme in the 1990s discussed as best-practice example, suggests that substantial change is possible. Like most past and current sustainability efforts, the Perspective project focused on individual behaviour change. It qualifies as best-practice example by demonstrating that living well, healthy and sustainably can go hand-in-hand. Marking the Dutch cycling culture as an example, Backhaus proposes that, rather than targeting individuals, future sustainable energy policies and programmes should best address infrastructures, social norms and collectives.

**Keywords** Dutch energy transition · Natural gas · Sustainable energy · Embodied energy · Energy research

J. Backhaus (✉)
ICIS, Maastricht University, Maastricht, The Netherlands
e-mail: j.backhaus@maastrichtuniversity.nl

© The Author(s) 2019                                                           71
F. Fahy et al. (eds.), *Energy Demand Challenges in Europe*,
https://doi.org/10.1007/978-3-030-20339-9_7

## INTRODUCTION

Today, heating, cooking and showering in Dutch households is largely fuelled by gas—not least due to the country's large natural gas reserves. Nevertheless, the Dutch government has committed to an energy system entirely based on renewables by 2050. This commitment entails that newly built homes are no longer required to be connected to the gas network, and gas sourcing companies have been ordered to cap, and within four years, completely stop extraction from a major gas field in the country's north after a series of increasingly severe earthquakes (Rijksoverheid 2018). Yet, while the country sets out to become 'gasless', the share of renewable energy is still very low and implementation is slow.

Developments in the Dutch energy sector have been, and will continue to be, strongly dependent on policies and trends in other countries, especially neighbouring countries in Europe's North-West. While the Netherlands is currently a net importer of energy, this is said to change from 2023 onwards according to the most recent National Energy Outlook (NEO) (Schoots et al. 2017). The 2017 NEO indicates around 5% share of renewables in total domestic energy use, which is projected to increase to 12% in 2020 and 17% in 2023. A steep downward trend of total energy consumption, especially in the built environment, observed between 2005 and 2016, is expected to continue. However, actual energy reduction in the built environment is considered to be much lower than theoretical calculations may suggest (Majcen et al. 2013), and demographic as well as socio-economic trends draw into question positive future projections (Brounen et al. 2012).

## SOCIO-MATERIAL DYNAMICS OF HOUSEHOLD ENERGY USE IN THE NETHERLANDS

With a share of over two-thirds, almost 90% of which is generated from gas, space heating consumes by far the most energy in the Dutch residential sector (Eurostat 2018). Research has shown that compared to households in other countries, Dutch households are less interested in energy-related home renovations including insulation or an upgrade of the heating system due to concerns regarding investment costs, among other reasons. Potentially a coping mechanism, but certainly of relevance in terms of comfort and social conventions, is a comparatively lower average indoor temperature (below 20 °C) at which people in the Netherlands feel comfortable at home (Kammerlander et al. 2014).

Energy use in the Dutch residential sector has decreased in recent years, and this downward trend is believed to continue, not least due to legislation particularly targeting the rental sector (Schoots et al. 2017). Cautionary voices are warning that the actual energy consumption of households in buildings with the most efficient energy labels (A–B) is higher than their theoretical energy consumption, which is used to inform energy policy (Majcen et al. 2013). Other research indicates that an ageing population and increasing wealth are likely to offset the effects of energy policies focusing on the physical and technical aspects of home energy use alone (Brounen et al. 2012). Since evidence suggests that the current and projected energy use of the Dutch dwelling stock and households is higher than assumed, additional energy saving measures are needed to achieve the full reduction potentials and meet energy consumption targets.

Personal mobility also makes up for a great share of energy use. More than 70% of Dutch households own at least one car (CBS 2017) and the total number of private vehicles on Dutch roads continues to rise (CBS 2019). Partially due to public policy, cars on Dutch streets are comparatively smaller and more efficient than in other countries in Europe's North-West. A particularity of the Netherlands is the Dutch 'cycling culture', which is catered to and supported by an extensive network of cycling paths as well as user-friendly bike rental and leasing schemes. In addition, private mobility needs are met by a well-maintained road infrastructure and a rather efficient public transportation system.

## Energy Policy in the Netherlands

Dutch national policy for the built environment focuses on energy efficiency, offering subsidies for heat pumps, and biomass, wood pellet or solar thermal heating systems to private homeowners, and since 2017 also to municipalities, provinces and public bodies. The 'energy efficiency you do now' (energie besparen doe je nu) programme provides cheap loans for energy efficiency renovations (e.g. insulation) to private homeowners and associations of apartment owners. Further, national government supports industry efforts with respect to electric heat pumps, district heating (geothermal and residual/waste heat) and the electrification of transport, including personal mobility. Furthermore, national energy policy requires utilities to support home energy savings, which often takes the form of energy saving tips, information on renovations and available subsidies, as well as smart meter-supported real-time data on home energy use. The overall goal for the national roll-out of smart meters is offering a smart meter to every Dutch household by 2020 and achieving at least 80% coverage. By the end of 2017, a total coverage of 16% had been achieved (author's calculation based on: CBS 2018; RVO 2018: 19).

Agreements have been made with municipalities and housing corporations that all rental apartments owned by housing corporations need to have an energy label of B or better, and all privately owned apartments need to have an energy label of C or better by 2020. Subsidies are available for apartment owners who either offer social housing or who plan rather ambitious energy efficiency renovations. An agreement with the construction, installation and energy sector states that every year, 300,000 existing flats are to be made energy efficient. For example, the block-for-block (blok voor blok) programme, running since 2012, comprises 14 projects targeting at least 33,500 flats that are, ideally, renovated cost efficiently, i.e. on block-level. In addition, the five biggest cities of the Netherlands together with the 32 middle to large-sized towns have developed a Smart City Strategy and six Dutch cities and industry partners are experimenting with 'smart city' concepts and technologies to achieve reductions in $CO_2$ emissions.

In recent times, interest has risen in supporting communities that are keen to take collective action by studying their business case (e.g. RVO) or by looking into the possibilities of an energy service company (ESCO) model for energy cooperatives (e.g. nmf Limburg). The Dutch cooperative sector is undergoing remarkable developments, currently fuelling 85,000 (1%) of Dutch households. Although there are no longer subsides for solar PV, energy cooperatives can profit from tax exemption schemes. In 2017, 100 new solar cooperatives have been established, leading to an increase of 53% compared to 2016 and a total solar capacity of 37 MW. Sixty-three of the new cooperatives benefit from the 'Postcode rose regulation' (Postcoderoosregeling), a national tax exemption scheme. This development will likely continue, with more than 200 projects planned for 2018. Although cooperative wind energy remained stable at 118 MW in 2017, a near doubling of capacity is expected for the period of 2018–2019 due to planned projects that emerged from close collaboration between several cooperatives, governments and commercial companies. Onshore wind energy is also increasingly cooperative-based, partly due to municipal requirements (HIER local energy monitor).

## Trends in National Household Energy Campaigns in the Netherlands

Sustainable Energy Consumption Initiatives (SECIs) in the Netherlands reflect Dutch energy policy in various ways (Table 7.1). The clear number one issue addressed is the energy efficiency of buildings and of appliances. The slow roll-out and uptake of energy efficiency measures, as well as renewable energy, has been recognised, and governmental actors at different levels have become better aligned by offering complementary support and services. While national energy policy for the built environment mainly addresses building envelopes, energy sources, (smart) systems and appliances, Dutch municipalities, and hence many SECIs, seek to facilitate and support uptake by collaborating with commercial actors, neighbourhood initiatives, cooperatives, etc. In addition, municipalities and environmental organisations try reaching individual households with more direct, tailored and accessible information about energy efficiency, available subsidies and other support schemes. For example, several SECIs consist of central information points, or energy coaches, who provide tailored advice. Another frequently found type of SECI provides energy efficiency equipment such as energy-efficient light bulbs, low-flow showerheads and draught-excluding tape to households for free; thus, aiming to inform about energy efficiency and providing an 'energy starter kit'.

**Table 7.1**  47 exemplary SECIs in the Netherlands according to their problem framing (Jensen et al. 2017)

| Problem framing | No. of initiatives |
|---|---|
| Changes in complex interactions | 2 |
| Changes in everyday life situations | 8 |
| Changes in individuals' behaviour | 29 |
| Changes in technology | 8 |

The good practice example below highlights the already high level of awareness of energy issues in the Dutch populace and indicates that—building on general awareness and willingness to change—personal, tailored advice paired with efficient technologies can help achieve significant consumption reduction of more than 40% in energy use.

## CASE STUDY: THE PERSPECTIVE PROJECT (PROJECT PERSPEKTIEF)

The Perspective project tested the possibility of living a low-energy lifestyle with a high level of well-being in a system of economic growth. The research project was carried out in the Netherlands from 1995 until 1998, financed by the then Ministry of Housing, Spatial Planning and the Environment (Ministerie van Volkshuisvesting, Ruimtelijke Ordening en Milieu; VROM). Supported by a research institute and two universities, the consultancy practitioner CEA implemented the project, involving private households. The initial 20 households were hand-picked and committed to aiming to reduce their energy use as much as possible over a period of two years. They were informed about the energy intensity of products and services, and coached and monitored throughout the entire run-time of the project towards achieving and maintaining a low-energy lifestyle (Project Team Perspective 1999).

In the Dutch context, the Perspective project is unique in terms of its focus, funding, duration and ambitions. National government was confronted with a constant increase of energy use, including by households. Research demonstrated that demand would continuously grow unless addressed. Therefore, the idea emerged to test whether energy use can be reduced and kept low, even if income increases, while well-being remains stable or improves. The Perspective project benefited from a general awareness of environmental and energy issues beyond the research and policy sphere. However, an exploration of what a reduction of indirect energy use means in practice was unprecedented (Schmidt 2017). A variety of households were selected for participation, and it appears that financial gains were as much a motivation to participate as possible environmental gains (Project Team Perspective 1999).

### Methods for Intervention

The methods of intervention addressing direct energy use were the provision of energy-efficient appliances, monitoring, information and coaching. At a kick-off event, participating households were informed about basic principles to reduce indirect energy use (e.g. quality rather than quantity, services rather than products). Second, people were given information on energy consumption and monitoring in print form. Most importantly, households received monthly coaching services. Advice focused on helping with the monitoring of home energy consumption, with thinking about saving strategies, with the planning of monetary spending, and with additional information on the energy intensity of products and services. A second event bringing together all participating households was organised about half-way through the project to thank and motivate everyone, exchange tips and experiences, and to commit people to the next period of maintaining reduced consumption levels. A final event was held to celebrate the successful completion of the entire undertaking.

Another method of intervention was the provision of 20% additional household income simulating economic growth. Households were obliged to follow a number of rules with respect to spending their increased income to ensure that their spending patterns would be following similar principles as before: no unusual donations or 'silly' expenditures; no more savings than prior to the project; and no big loans and investments. Any purchase costing more than 500 Dutch guilders (approx. €225) had to be discussed with the coach who then gave advice based on potential energy impact.

### Framing the Energy Challenge

The project focused on households who already lived in energy-efficient homes and who were willing to commit to a two-year, longitudinal study on saving embodied (indirect) and direct energy use. As the main goal was achieving maximum energy savings with the help of (or despite) a 20% increase in household income, energy and energy savings were framed as something positive for all household members, in terms of health or time efficiency, as well as for the environment. The project created several changes in everyday life situations regarding information, support and finances, to evaluate the possibility of maintaining a low-carbon lifestyle in a scenario of economic growth and increased personal wealth.

## *Outcomes and Outputs*

The goal of the Perspective project was a 40% reduction in energy use compared to similar households in less efficient homes. An average of 43% of reduction was accomplished, about half resulting from reductions in direct, and half from indirect energy use. Miniscule monitoring was done through meter readings as well as the careful registering of all products bought, and services used. The registering of product purchases was made somewhat easier by means of a self-learning system with barcode reader, which required the manual entering of data only the first time an item from the supermarket was scanned. All subsequent scans were then automatically registered. Some products had to be weighed in the monitoring process. In addition, interviews were conducted to gain insights into people's emotions and experiences, for example with respect to comfort and well-being and the value they attach to different consumption categories. Personal coaches took note of dilemmas people faced, such as the desire of wanting to go to a faraway place for vacation and, due to their commitment to the project, the requirement to take a low-energy holiday instead.

Overall, households reached the target by reducing their direct and indirect energy use and—simulating potential effects of economic growth on households—increasing their level of spending and well-being. They achieved a reduction in energy use in all categories measured: transport, food, living, hygiene, clothing and leisure. Monetary spending increased in the categories of food and living and decreased in the categories of direct energy use and leisure (Project Team Perspective 1999).

## CONCLUSION

The catalogue of exemplary SECIs presented in this chapter did not aim to be representative of all sustainable energy initiatives in the Netherlands, yet a heavy focus on technological solutions and individual behaviour change can be seen in the figures shown in Table 7.1. Initiatives focusing on changing collective conventions for significant reductions in energy use (as tested by the ENERGISE project) are still rare, yet crucial to meet national renewable energy and energy efficiency targets. An important finding of the presented best practice project Perspective, and many other projects, is that living healthy and well is possible with a smaller footprint. A potential improvement in the design and approach of the Perspective research project, that successfully provided individual coaching services to participating households, could have involved more interaction between participants to share ideas and experiences, and to tackle social conventions collectively.

To achieve the country's contribution to meeting targets of the Paris Agreement, current positive trends in the Netherlands need to be maintained and their pace and reach needs to increase. The 100% goal for renewable energy by 2050 appears steep, and seems to require the more forceful pursuance of currently still nascent cooperative models. Similarly, the downward trend of energy use in the built environment needs to continue, accelerate and potentially be enjoyed with caution as the discrepancy between potential and actual reduction is unknown. In terms of renewable energy, recognising that incumbent actors are reacting slowly at best, policy actors and supporting agencies have recently started to develop better support measures for local, community-based renewable energy initiatives. In terms of energy efficiency, policy and other national actors seem to have hardly taken note of the role and relevance of collective conventions that might require more collective models of engagement. Hopefully, in the future, the importance of challenging unsustainable social norms that are enabled by and entrenched in existing and even in emerging infrastructures is recognised. Such recognition could result in more sustainable policies, initiatives and infrastructure investments.

**Acknowledgements** The author would like to thank Jochem Pennekamp for his tireless efforts in seeking out more than 40 SECIs in the Netherlands that have informed this study.

# References

Brounen, D., Kok, N., & Quigley, J. M. (2012). Residential energy use and conservation: Economics and demographics. *European Economic Review, 56*(5). https://doi.org/10.1016/j.euroecorev.2012.02.007.

CBS—Centraal Bureau voor de Statistiek. (2017). *Huishoudens in bezit van auto of motor; huishoudkenmerken, 2010–2015.* Available at https://statline.cbs.nl/StatWeb/publication/?DM=SLNL&PA=81845ned.

CBS—Centraal Bureau voor de Statistiek. (2018). *Huishoudens; samenstelling, grootte, regio, 1 januari.* Available at https://statline.cbs.nl/statweb/publication/?vw=t&dm=slnl&pa=71486ned&d1=0-2,23-26&d2=0&d3=0,5-16&d4=(l-1)-l&hd=090402-0910&hdr=t,g3&stb=g1,g2.

CBS—Centraal Bureau voor de Statistiek. (2019). *Personenauto's; voertuigkenmerken, regio's, 1 januari.* Available at https://opendata.cbs.nl/statline/#/CBS/nl/dataset/71405ned/table?dl=9FAA.

Eurostat. (2018). *Energy consumption in households.* Available at https://ec.europa.eu/eurostat/statistics-explained/index.php/Energy_consumption_in_households.

HIER opgewekt, *Local Energy Monitor.* Available at https://www.hieropgewekt.nl/local-energy-monitor.

Jensen et al. (2017). *Comprehensive open access dataset of Sustainable Energy Consumption Initiatives (SECIs).* ENERGISE—European Network for Research, Good Practice and Innovation for Sustainable Energy, Grant Agreement No. 727642, Deliverable No. 2.3.

Kammerlander, M. et al. (2014). *Individual behavioural barriers to resource-efficiency.* POLFREE—Policy options for a resource-efficient economy. Grant Agreement No. 308371, Deliverable 1.6

Majcen, D., Itard, L. C. M., & Visscher, H. (2013). Theoretical vs. actual energy consumption of labelled dwellings in the Netherlands: Discrepancies and policy implications. *Energy Policy, 54.* https://doi.org/10.1016/j.enpol.2012.11.008.

Project Team Perspective. (1999). *Minder energiegebruik door een andere leefstijl?* Project Perspectief, December 1995–Juni 1998. Eindrapportage. VROM.

Rijksoverheid. (2018). *Kabinet: einde aan gaswinning in Groningen; Nieuwsbericht* 29 March 2018. Available at https://www.rijksoverheid.nl/actueel/nieuws/2018/03/29/kabinet-einde-aan-gaswinning-in-groningen.

RVO—Rijksdienst voor Ondernemend Nederland. (2018). *Marktbarometer Aanbieding Slimme Meters. Voortgangsrapportage 2017.* Available at https://www.rijksoverheid.nl/documenten/rapporten/2018/06/19/voortgangsrapportage-marktbarometer-aanbieding-slimme-meters-2017.

Schmidt, T., Project leader of the project Perspektief, personal communication on 27 June 2017, 8.45–9.30 am.

Schoots, K., Hekkenberg, M., & Hammingh, P. (2017). *National energy outlook—Summary.* Available at https://www.ecn.nl/nl/energieverkenning/.

# CHAPTER 8

# The Energy Challenge in Hungary: A Need for More Complex Approaches

*Edina Vadovics*

**Abstract** Vadovics first describes the motivations behind and objectives for energy policy in Hungary. This is followed by an overview of sustainable household energy consumption initiatives, and their classification according to the ENERGISE problem framing typology. It is shown that although the initiatives are very diverse, they are dominated by those focusing on individual behaviour and technology change. Then, one of the initiatives, a local climate club, is introduced and analyzed. It is described how complex initiatives, including small group-based ones, can support change towards, and create capacities for, more sustainable energy use. The chapter then concludes with policy conclusions relevant to both Hungary and Europe.

**Keywords** Energy policy · Sustainable energy consumption · Problem framing · Behaviour change · Small groups

E. Vadovics (✉)
GreenDependent Institute, Gödöllő, Hungary
e-mail: edina@greendependent.org

© The Author(s) 2019
F. Fahy et al. (eds.), *Energy Demand Challenges in Europe*,
https://doi.org/10.1007/978-3-030-20339-9_8

83

# INTRODUCTION

Hungary is poor in fossil fuel resources but at the same time close to 90% of its total primary energy supply comes from fossil fuel and nuclear sources (MEKH 2017). Thus, dependence on external fossil fuel and non-renewable resources is one of the most important issues facing energy and climate policy-makers. With regards to energy system ownership structures, there is an explicit government policy to establish a state-owned, centralised infrastructure as the main means for the provision of energy for the household sector, one of the largest final energy user sectors with 31%, followed by the transportation (22%) and industrial (21%) sectors (MEKH 2017). Perhaps it is not surprising that Hungary is lagging behind other European countries in terms of both renewable energy utilisation and community energy, as well as supporting the transition to a prosumer culture, all of which would require a more flexible and less centralised energy system. At the same time, per capita carbon emissions in Hungary are lower than the European average, and, in fact, lower than in most European countries (EEA 2019, based on data from 2016). This fact if considered together with the rather high (cc. 40%) saving potential in the household sector means that there is a so far unrealised potential towards a low-carbon economy (Energiaklub 2011). Although, according to the latest Eurobarometer survey (2018), climate change is not considered to be a central issue by Hungarian citizens, it is worth noting that there is a higher than average (76%) support for the common European energy policy (Bart et al. 2018).

## SOCIO-MATERIAL DYNAMICS OF HOUSEHOLD ENERGY USE IN HUNGARY

With the household sector taking up the largest share of final energy use (31%), building energy efficiency is an important factor in energy and climate policy. This is especially so given the fact that Hungary's building stock is technically obsolete, with a large proportion of buildings lacking adequate insulation and/or energy-efficient heating systems (Bart et al. 2018). About 86% of homes are owned privately, and the share of the population living in detached houses is relatively high, 63% (Eurostat 2018). These households often use a mix of fuels for heating, typically natural gas and wood. About 30% of the population live in flats, a considerable proportion of which are blocks of flats built using industrial technologies (Eurostat 2018; Jensen et al. 2018). The use of district heating and other joint/community solutions are hindered by negative social attitudes towards public or joint ownership schemes (Jensen et al. 2018).

A considerable share of society (around 35%) live under the 'subsistence' levels and 21% live in fuel poverty (Fülöp and Lehoczki-Krsjak 2014). 27% of homes have inadequate walls and roofs, and 9% of the population are unable to keep their homes warm (Eurostat in HBS–FoE Hungary 2018). Within the European Union, Hungary has one of the highest rates of housing deprivation (Eurostat 2018). Thus, the affordability of energy is a major issue and the popular policy of the government is to regulate the price of energy.

As for the use of energy, the level of consciousness is low. The majority of the households do not follow their energy consumption data and the household appliances stock is inefficient on a large scale (Slezák et al. 2015). On the other hand, in a survey the Hungarian population expressed willingness and interest in energy-efficient home improvements (Fülöp and Kun 2014), but, on the whole, households lack the financial resources to act on this interest.

## Energy Policy in Hungary

The most important goals of Hungarian energy policy (MND 2012) are the provision of affordable energy, long-term sustainability, supply security and economic competitiveness. Special emphasis is placed on tackling the energy dependency of the country by means of: (i) energy savings; (ii) increasing the share of renewable energy sources; (iii) safe nuclear energy and the electrification of transport based on this; (iv) creating a bipolar agriculture (food production and energy-geared biomass production); and (v) better integration to the European energy infrastructures (Jensen et al. 2018).

A characteristic feature of recent energy policy is the pivotal role of the government. In this context, important measures were taken by the government such as setting up a 100% state-owned National Public Utility Company (next to the private utility companies operating in the country) to ensure the security of energy supply. It is an explicit policy of the government to keep energy prices low. Related measures have included the appointment of a governmental commissioner and the regulation of utility prices for the household sector (Jensen et al. 2018).

In the context of energy saving and efficiency, the main focus is put on the household sector and the building stock. However, during the last few years, relevant policy support has been volatile (e.g., relevant policies had been announced and then re-called), sporadic, short-term with resources running out in days, and actual incentives have targeted the public rather than the household sector (Bart et al. 2018; HBS–FoE Hungary 2018; Jensen et al. 2018).

There are several gaps and controversies in recent Hungarian energy and climate policy that give rise to discussions among experts in the sector. It is claimed that the potential for energy saving is higher than that predicted by the government. Thus, national plans are not ambitious enough (Slezák et al. 2015; Bart et al. 2018). Additionally, the expected growth rate of energy consumption is disputed, and alternative estimations argue that the planned capacity enlargement of the Paks Nuclear Power Plant is not necessary in view of the country's potential for energy efficiency improvements (Lechtenböhmer et al. 2016). There is also a debate about the regulation of household energy prices, especially as it took place without any differentiation based on income levels, housing deprivation status, etc. (Jensen et al. 2018).

Finally, it needs to be noted that in line with EU requirements, the process of developing the Integrated National Energy and Climate Plan has begun in Hungary. It is expected that the Plan, once adopted, may bring some changes in the policy context.

## TRENDS IN HOUSEHOLD ENERGY CAMPAIGNS IN HUNGARY

Governmental energy-related campaigns have been dominated in the last years by the 'war on utility costs'—an overall populist price policy of the government (Slezák et al. 2015). To a lesser degree, in line with the Energy- and Climate Awareness-Raising Action Plan of Hungary, policies have been supported by certain awareness-raising activities.

Regarding monetary incentives, the main financial instrument managed by the central government to promote investments aimed at furthering energy efficiency in households has been the so-called 'Warmth of the Home Programme' grant scheme. Set up in 2014, it has provided financial support in several phases. For instance, grants are available for the replacement of inefficient household appliances, inefficient doors and windows, etc. However, the available funding has been insufficient and has always been sourced out within days, indicating a high level of interest from the population (HBS–FoE Hungary 2018; Jensen et al. 2018).

In addition, interest-free soft loans for energy efficiency improvements are available both for the household and the SME sectors (Jensen et al. 2018). Some municipal governments also provide incentives for energy efficiency renovations for households, especially for those living in apartment blocks (Slezák et al. 2015).

## SUSTAINABLE ENERGY CONSUMPTION INITIATIVES (SECIs) IN HUNGARY

A recent review conducted in the framework of the European ENERGISE project revealed that in line with findings for all European countries (see Chapter 2), Hungarian SECIs are dominated by those focusing on individual behaviour and technology change (see Table 8.1). Nonetheless, the Hungarian SECIs reviewed in the ENERGISE project are very diverse in terms of their objectives, target groups, the type of organisations implementing them, the methods they apply, their funders, etc. Some SECIs are directly impacted by the energy policies of Hungary and the European Union, as support and funding available for them is determined by policy objectives. Other SECIs are specifically created to aid the implementation of policies, or prepare various stakeholders for the implementation of a particular policy. For example, projects have been carried out to educate stakeholders about new energy regulations and building directives. There are, also, less technology-oriented projects

Table 8.1 Number of national SECIs according to their problem framing in Hungary

| Problem framing | No. of initiatives |
| --- | --- |
| Changes in Complex Interactions | 6 |
| Changes in Everyday Life Situations | 4 |
| Changes in Individuals' Behaviour | 15 |
| Changes in Technology | 20 |

in this category, for example, those that are intended to bring about attitude and behaviour change to prepare the general population for the impacts of climate change and to motivate more sustainable energy use patterns.

On the other hand, SECIs often respond to needs that are not met by policy, and aim to go beyond what is requested by policies, or even challenge policies. Consequently, these kinds of SECIs are often funded from sources other than the national government: from private foundations, companies or local municipalities, from European frameworks, and, still less often, from community funding (e.g., crowdfunding). Examples of these types of SECIs include the *Biomass briquettes programme* created to provide an alternative, environmentally friendly fuel source as well as employment opportunities for a community living in energy poverty, or the very innovative *Climate ticket* (*Klímajegy*) initiative that made it possible for individuals and organisations in a specific region to voluntary compensate their carbon footprint through supporting local sustainable energy projects (e.g., installing solar panels, planting trees).

Both of these categories include initiatives that correspond to each of the four problem framings presented in Table 8.1. Below, we introduce an initiative that was categorised as one that brings about 'changes in complex interactions'.

## CASE STUDY: GÖDÖLLŐ CLIMATE CLUB

The Climate Club was established in 2009 by GreenDependent Association in order to raise awareness of climate change issues in households, establish links between climate change and household consumption, and create a sense of responsibility for consumption and lifestyle-related emissions in households. Climate Club[1] members live in or around Gödöllő, a town in Central Hungary. The Club was formed as part of a European research project called Changing Behaviour.[2] Thus, it started as a pilot project, but is ongoing to this day (February 2019).

[1] More information about the Gödöllő Climate Club can be found at http://klimaklub. greendependent.org/en.html.

[2] To learn more about the Changing Behaviour project, please visit http://www.energy-change.info/.

The core activity of the Club is its monthly meetings where members discuss, in an informal setting, climate change, energy-related and environmental issues, ideas and concerns. Alternatively, Club members invite experts to have a discussion or give a presentation on a given topic of interest.

## Methods for Intervention

The Changing Behaviour research project studied successful and less successful demand-side management programmes in an effort to establish general success factors (Mourik et al. 2009; Vadovics and Boza-Kiss 2013). As a partner in the project, GreenDependent attempted to incorporate many of the identified success factors in the intervention methodology in order to create lasting change. For example, creating a community is important so that participants do not feel alone in their efforts; besides, they can share experience and learn from each other, too. Some methods and tools were developed to allow for flexibility as well as to cater to the communication and learning needs of people with different backgrounds and various levels of experience.

Building on the findings of the Changing Behaviour project (Heiskanen et al. 2010), Table 8.2 summarises how small groups, and, in particular, the Climate Club, can help overcome barriers to behaviour change and, at the same time, create capacities and skills for change.

Households were not involved in the design phase of the pilot project, as it was based on 'best practice methodology' identified in the Changing Behaviour project. Apart from this, however, they are invited to take an active role in planning the content of the monthly meetings, in initiating activities in the larger community, and in the organisation of the club meetings, e.g., by providing homemade food and drinks.

## Framing the Energy Challenge

The energy and climate change challenge has been communicated to and discussed with households as a challenge that potentially has an impact on all aspects of their lives, which, in turn, has an impact on how serious the challenge is going to be. Households were invited to consider their everyday life from the point of view of how much energy they use and what they use it for, taking a systemic perspective. They also considered how they can limit the impact of energy and climate change in

**Table 8.2**  Ways in which small groups can help overcome barriers to behaviour change

| Capacities | Description | Barrier to behaviour change | How the Climate Club can help overcome barrier |
|---|---|---|---|
| Personal | Individuals' understanding of the issue, their willingness and ability to act, their values skills and enthusiasm | Lack of knowledge and understanding, lack of willingness and skills, helplessness | Sharing and creating knowledge<br>Providing advice, skills, motivation and encouragement<br>Members can see that 'others are also doing their bit'<br>Assurance that being 'green' is normal |
| Infrastructural | Facilities and structures enabling sustainable living available in the community | Current socio-technical infrastructures | Creating knowledge network on the carbon intensity of lifestyles and the low-carbon solutions available in the community |
| Organisational | Values held by formal organisations in the community | Social conventions, helplessness | Challenging existing institutions<br>Changing taken-for-granted beliefs about modern life and creating a supportive environment for problematising current lifestyles |
| Cultural | Legitimacy of sustainability and low-carbon living in the community | Social dilemmas, helplessness | Creating a community of individuals prepared to change their lifestyles and promote these changes to others and by doing so creating legitimacy for sustainable and low-carbon values and living |

*Source* Vadovics and Heiskanen (2010)

their individual homes and everyday routines by changing their practices and behaviour, and by making technological changes, e.g., through energy-efficient home improvements. Since they were part of the group, they were encouraged to discover and consider the social aspects of energy use, e.g., the conventions and expectations that govern their everyday practices, the things they can influence and can perhaps change together, and the local opportunities, knowledge and skills available for making more sustainable energy use possible. After spending some time together as members of the same Club, they became motivated to get engaged in the local community, and organise awareness-raising activities at community events. The Climate Club also engages in areas more indirectly related to energy consumption, such as producing food, dealing with waste, and so on. The Club works with other NGOs in the town and has been networking with similar initiatives in Hungary.

### Outcomes, Successes of the Initiative

Because of its more informal nature, there was no comprehensive study done on the carbon footprint reduction or energy saving achieved by the Gödöllő Climate Club members, but there are indications that most members achieved at least 10% reduction in energy use since they joined the Club. More importantly, the Climate Club has become a small group of dedicated individuals who appreciate the additional knowledge and the sense of community as a primary value provided by the monthly meetings. It is clear that most members feel closely associated with the group, and have a feeling of ownership, which seems to be increasing with time (Vadovics and Boza-Kiss 2013). Overall, the initiative is considered to be a success story, with members continuing to meet to discuss environmental and energy issues, as well as to take action.

### CONCLUSION

The example of the *Gödöllő Climate Club* is an interesting SECI for various reasons. Firstly, it was conceived as a pilot project in a European research project. The methodology that was piloted and evaluated in the project was used later for the development of larger (national and international) projects. Secondly, since the Climate Club is still active after nine-ten years, it is a good example of how a pilot project can turn into a continuous project, partly run by the community, and partly by the organisation that piloted it originally. Thus, an important lesson learnt is that well-planned projects with impact can continue successfully beyond the pilot and fully funded stage even though this continuation was entirely dependent on the organisation managing it and the local community, as there was no support provided at either the national or European level.

Finally, although the *Gödöllő Climate Club* is an example of how a SECI can be designed to create change in complex relations, the majority (more than 75%) of the SECIs reviewed for Hungary focus on individual behavioural change and technological aspects of sustainable energy consumption (see Table 8.1). These are important factors, and need to be tackled in initiatives, but there is need to integrate them with social considerations, such as how energy is used, what it is used for, and the communities within which it is used. Since policies have a great influence on what kind of SECIs are implemented and how change is understood to happen, there is also need at that level to take the social aspects of energy consumption into consideration.

**Acknowledgements** I would like to express my gratitude to colleagues who have contributed in various ways to previous versions of this paper. To József Slezák, who has provided input that greatly assisted me in drafting the sections related to energy policy in this paper, and to Gergő Horváth, Andrea Király, Szandra Szomor and Kristóf Vadovics who provided additional support with background research and data collection.

## References

Bart, I., Csernus, D., & Sáfián, F. (2018). *Analysis of climate-energy policies & implementation in Hungary*. National Society of Conservationists–Friends of the Earth Hungary.

Energiaklub. (2011). *Negajoule 2020—Energy efficiency potential of Hungarian residential buildings*. Budapest, Hungary.

EEA (European Environment Agency). (2019). *EEA greenhouse gas—Data viewer*. Available at https://www.eea.europa.eu/data-and-maps/data/data-viewers/greenhouse-gases-viewer. Last accessed 11 February 2019.

Eurostat. (2018). *Housing statistics*. Based on data extracted May 2018. Available at https://ec.europa.eu/eurostat/statistics-explained/index.php/Housing_statistics#Tenure_status. Last accessed 28 February 2019.

Fülöp, O., & Kun, Zs. (2014). *Lakossági Energiahatékonysági Barométer 2014.* ENERGIAKLUB Szakpolitikai Intézet és Módszertani Központ, Budapest, Hungary.

Fülöp, O., & Lehoczki-Krsjak, A. (2014). Energiaszegénység Magyarországon. *Statisztikai Szemle, 92,* 8–9.

Heinrich Böll Stiftung and National Society of Conservationists–Friends of the Earth Hungary (HBS–FoE Hungary). (2018). *Energia Atlasz 2018. Tények a megújuló energiaforrásokról Európában.*

Heiskanen, E., Johnson, M., Robinson, S., Vadovics, E., & Saastamoinen, M. (2010). Low-carbon communities as a context for individual behavioural change. *Energy Policy, 38*(12), 7586–7595.

Jensen, C., et al. (2018). *30 national summary briefs of national energy supply and demand.* ENERGISE—European Network for Research, Good Practice and Innovation for Sustainable Energy, Grant Agreement No. 727642, Deliverable 2.5. Specific chapter on Hungary available at http://www.energise-project.eu/sites/default/files/content/D2.5_Hungary.pdf. Last accessed 28 February 2019.

Lechtenböhmer, S., Prantner, M., Schneider, C., Fülöp, O., & Sáfián, F. (2016). *Alternative and sustainable energy scenarios for Hungary.* Budapest, Hungary: Zöld Műhely Alapítvány.

MEKH (Hungarian Energy and Public Utility Regulatory Authority). (2017). *Hungarian Energy Balance 2017.*

MND. (2012). Nemzeti Fejlesztési Minisztérium/Ministry of National Development (MND). *National Energy Strategy 2030.*

Mourik, R. M., et al. (2009) *Past 10 year of best and bad practices in demand management: A meta analysis of 27 case studies focusing on conditions explaining success and failure of demand-side management programmes.* Deliverable 4 of the Changing Behaviour project.

Slezák, J., Vadovics, E., Trotta, G., & Lorek, S. (2015). Consumers and energy efficiency—Country report for Hungary. An inventory of policies, business and civil initiatives at the national level, focusing on heating, hot water and the use of electricity. December 2015. *EUFORIE—European Futures for Energy Efficiency.*

Vadovics, E., & Boza-Kiss, B. (2013, June). Voluntary consumption reduction—Experience from three consecutive residential programmes in Hungary. Residential energy master as a new carrier? In *Proceedings of the SCORAI Workshop,* Istanbul.

Vadovics, E., & Heiskanen, E. (2010, October 26–29). *Understanding and enhancing the contribution of low-carbon communities to more sustainable lifestyles: The case of the Gödöllő Climate Club in Hungary.* Poster presented at the ERSCP-EMSU conference in Delft, Holland.

# CHAPTER 9

# Slovenia: Focus on Energy Efficiency, Community Energy Projects and Energy Poverty

*Lidija Živčič and Tomislav Tkalec*

**Abstract** Slovenia has a small energy sector, where oil (45%) represents the main energy source. Electricity generation is equally divided between hydropower, nuclear energy and coal. Trends in energy policy go in the direction of maintaining status quo. A significant percentage of households live in energy poverty due to combination of low energy efficiency of buildings, high ownership rates, and low incomes. The Sustainable Energy Consumption Initiatives (SECIs) reviewed in this chapter are generally ahead of the trends in national energy policies. The progressive nature of many SECIs is evident in the field of energy efficiency and diversity of effective approaches, in particular in the cases of community renewable energy initiatives and the problem of energy poverty. Policymakers still do not fully appreciate the relevance of these areas for a sustainable transition. Especially in areas of community energy and energy

L. Živčič (✉) · T. Tkalec
Focus Association for Sustainable Development, Ljubljana, Slovenia
e-mail: lidija@focus.si

T. Tkalec
e-mail: tomi@focus.si

© The Author(s) 2019
F. Fahy et al. (eds.), *Energy Demand Challenges in Europe*,
https://doi.org/10.1007/978-3-030-20339-9_9

poverty, SECIs provide recommendations to decision-makers on how to proceed in dealing with these issues.

**Keywords**  Slovenia · Energy poverty · Community energy projects · Energy policy · Sustainable energy initiatives

## INTRODUCTION

Slovenia has a small energy sector, with final energy consumption in 2017 of 4.92 Mtoe (57,242 GWh). Oil (45%) is the main energy source, followed by electricity (24%), renewables (14%), natural gas (12%), heat (4%) and solid fuels (1%) (SURS 2018). Electricity generation can be divided into three parts—hydropower (38%), nuclear power (37%) and thermal power (22%)—that vary slightly from year to year because of weather conditions and the amount of rainfall, which influences generation in hydropower plants. The biggest share of thermal power comes from one coal-fired plant (lignite). Renewables contribute only a small share of electricity generation, with solar accounting for less than 2%, and even less wind energy (0.02%) (ARSO 2015). Slovenian energy independence in 2017 was 52%.

Even though the EU is one of the most developed areas in the world, between 50 and 125 million EU citizens are estimated to be energy poor. The situation is most severe in the Eastern Europe Member States, including Slovenia. In the majority of the new Member States, up to 30%, or even more, of households are struggling with energy poverty. Following a brief overview of household energy use and energy policy in Slovenia, this chapter presents details of a sustainable energy initiative that aimed to alleviate energy poverty in Slovenia and neighbouring countries.

## SOCIO-MATERIAL DYNAMICS OF HOUSEHOLD ENERGY USE IN SLOVENIA

Future growth in electricity consumption must be considered given the expansion of air conditioning and electric vehicles. Energy efficiency measures, in particular insulation of multi-apartment buildings and family houses, serve as a counterbalance to these increases. However, improvements in thermal insulation are usually concentrated in the better-off sectors of the population, while the less well-off are less able to invest in improving energy efficiency of their dwellings. A significant percentage of households live in energy poverty, primarily because of low energy efficiency of buildings in combination with high ownership rates (more than 95% of people live in their own flat or house) and low incomes. Due to rising prices of heating oil in the last 10 years, there is a tendency to replace oil for heating with cheaper alternatives. While some households use heat pumps and gas, an increasing number choose biomass, as wood is the cheapest option. However, the large percentage of heating with biomass leads to an increasing problem with air pollution.

## ENERGY POLICY IN SLOVENIA

The government has worked on a new national energy concept since the autumn of 2014, but the programme has not yet been adopted. The current version focuses on keeping the status quo and preparing for changes in the longer term. Slovenia's vision for the energy sector is gradually to transition to low-carbon energy sources by focusing on efficient energy consumption, use of renewable energy sources (RES) and the development of active electricity-distribution networks. This strategy will likely envisage a strong reliance on nuclear energy and further development of hydroelectric power (Export.gov 2017).

While the gas and oil markets are somewhat privatised, electricity production is still in state hands. The major electricity producers in Slovenia, Holding Slovenske Elektrarne (HSE) and Gen Energija (GEN), are fully owned by the state of Slovenia. These two companies formed a single unit until 2001. In recent years, the government has considered reverting to a single company through the merger of HSE and GEN, as HSE did not have the necessary financing for the construction of a new coal-fired generator at the Šoštanj Thermal Power Plant (TEŠ 6). Nonetheless, construction on the TEŠ 6 project continued despite concerns about its cost, commercial feasibility, environmental impact and the perceived lack of transparency surrounding the project. Despite these concerns, the government of Slovenia provided the necessary loan guarantees to finish the project, and TEŠ 6 went on-line in 2015 (Export.gov 2017). There is still no definite answer about the timing of closure of the coal power plant TEŠ, as its life expectancy until 2054 is not likely to be met because of economic and environmental reasons.

RES are still not a government priority, and bigger investments (apart from hydropower) are not planned until after the year 2030. Increased hydroelectric power generation is one of the strategic objectives of the government's energy policy. Further upgrading of existing stations on the Sava and Drava rivers is planned, as well as the construction of several new plants on the Sava, and several other small hydroelectric power plants. Together with the new plants, these renovations should create an additional 470 MW of hydroelectric capacity in the near future (Export. gov 2017).

GEN Energija has prepared a plan for a second nuclear production facility. However, the government's decision on the timing of any possible nuclear expansion will depend on energy needs, available financing and public sentiment about nuclear energy. GEN Energija currently owns half of the Krško Nuclear Power Plant, which is co-owned by Croatia's state company HEP (Export.gov 2017).

Energy efficiency and energy refurbishment of the building stock are perceived as priority measures, but are not fully visible in government financial schemes and policies. Energy communities are only slowly entering the discourse.

## TRENDS IN NATIONAL HOUSEHOLD ENERGY CAMPAIGNS IN SLOVENIA

National campaigns in Slovenia are run mainly through the national organisations Eco Fund and Energy Agency. Eco Fund has programmes and financial aids for energy efficiency (EE) and RES measures (e.g. energy refurbishment of buildings, replacement of old inefficient heating systems, energy advising, energy poverty alleviation programmes, co-financing investments in RES, subsidies for electric cars). Energy Agency is responsible for tenders for support schemes for RES projects.

The Ministry for Environment is also active in campaigns for cleaner air, targeting air pollution from wood burning. Other non-governmental stakeholders and actors, including utility companies, run campaigns on RES projects and 'prosumership' of RES electricity through a net metering scheme, which allows consumers who generate their own electricity to use that electricity at any time. Community groups run campaigns based on community (RES) projects, energy efficiency, energy poverty and sustainable mobility. Apart from being active in the promotion of EE and RES, community groups are also involved in campaigns against fossil fuels and nuclear power.

## CHARACTERISTICS OF NATIONAL SECIs IN SLOVENIA

A review of recent Sustainable Energy Consumption Initiatives (SECIs) in Slovenia, according to their problem framing, reflects the dominance of traditional problem framings that prioritise changes in individuals' behaviour and changes in technology (Table 9.1). There are only a few initiatives that understand the challenge of changing energy use as a more complex and collective concern and several initiatives that target changes in everyday life situations.

**Table 9.1** Number of national SECIs in Slovenia according to their problem framing

| Problem framing | No. of initiatives |
| --- | --- |
| Changes in complex interactions | 4 |
| Changes in everyday life situations | 7 |
| Changes in individuals' behaviour | 23 |
| Changes in technology | 11 |

In Slovenia, SECIs are generally ahead of the trends in national energy policies. The progressive nature of many SECIs is already evident in the field of energy efficiency and diversity of effective approaches, in particular in the cases of RES initiatives, community energy projects and the problem of energy poverty. Policy-makers still do not fully appreciate the relevance of these areas for a sustainable transition. Especially in areas of community energy and energy poverty, SECIs provide examples to decision-makers on how to proceed in dealing with these issues.

Often, non-governmental actors take the initiative in these areas and, on the basis of their acquired knowledge and experience, influence the policy-makers who, due to these SECIs, are beginning to deal with the topics concerned (namely energy poverty and community energy).

There is some attention paid in SECIs to the socio-material specifics of energy use. Energy efficiency is high on the agenda of several SECIs. Energy poverty is highlighted as one of the socio-material aspects and is reflected in ten of the identified SECIs. One visible characteristic of SECIs that target energy poverty is that many of them work with such households in a variety of manners, from working towards energy retro-fits, to providing home audits, energy advising, awareness raising, under-standing of energy and heating bills, participatory workshops on energy saving, providing financial support and other support measures.

The majority of identified SECIs focus on changes in individuals' behaviour (23), followed by changes in technology (11), while those that focus on changes in complex interactions are scarce (4). However, there are seven SECIs focusing on changes in everyday life situations. The majority of SECIs identified in Slovenia are run at a cross-national level (24), 15 are implemented at a national level, 5 at a regional level and 5 at a local level. The majority of the Slovenian cases are built around energy efficiency (combination of reduction and substitution) and substitution.

Governmental programmes are rather scarce, and many initiatives come from EU projects, energy agencies and initiatives by the envi-ronmental NGOs. Actions are mostly not targeted to specific socio-demographic groups, although there are quite a high number of initi-atives targeting low-income households (10), which shows that energy poverty is recognised as an important issue.

## CASE STUDY: REACH (REDUCE ENERGY USE AND CHANGE HABITS)

REACH (Reduce energy use and change habits) is an IEE-funded pro-ject (2014–2017) aimed at reducing energy consumption in low-income (energy poor) households. It was implemented in Croatia, Macedonia, Bulgaria and Slovenia. In Slovenia, the project was administered by FOCUS. In all countries, practical activities of the project—energy advis-ing in households—were conducted on a regional level. In Slovenia, the project was run in the Pomurje and Zasavje regions.

REACH built on the success of a previous IEE project, ACHIEVE. Project REACH had two overall objectives: (i) to empower energy-poor households to take actions to save energy and change their habits; and (ii) to establish energy poverty as an issue that demands tailor-made policies and measures at local, national and EU level. The specific aims of the project were:

- To compile data and analyse energy poverty situations in four countries in order to form definition(s) of energy poverty and policy recommendations.
- To engage and empower local actors to tackle energy poverty.
- To empower energy-poor households to reduce their energy and water use and to provide some of them with further support for tackling their problems.
- To engage decision-makers in tackling energy poverty as an issue that demands structural tailor-made solutions, to provide them with recommendations for addressing the problem and to create a platform for concerted formulation of structural solutions at the national and EU level.

### *Framing the Energy Challenge*

Energy poverty in Slovenia is becoming an increasing problem as rising energy prices surpass the rise of incomes of the population. Thus, the expenditure on energy for households in the first income quintile has risen sharply in the recent couple of years, representing 17% of all available resources of individual households in 2010 (in 2000, this share was 13%). In the context of EU policies, the issue of energy poverty is becoming more and more visible, but there is no single definition of who is energy poor. Despite the lack of definition, energy poverty is being tackled by some policies: governmental analysis of energy poverty from 2010 highlights energy poverty as a rising issue, and the National Energy Action Plan 2020 and the Operative programme 2014–2020 list energy retrofit of energy-poor households as an objective. Hence, some measures for addressing energy poverty already exist in Slovenia: the national programme for visiting energy-poor households by advisers of the national energy efficiency advising network, support for energy retrofits of energy-poor households (100% subsidy), and support for the replacement of heating systems in energy-poor households in areas that are particularly burdened with air pollution. However, further steps are necessary to address the problem fully.

## Methods for Intervention

The first step of the project was to map the local and national situation in the field of energy poverty. National analyses of the energy poverty situation were conducted in order to (i) gain insight into the situation; (ii) provide a basis for fine-tuning the action; and (iii) provide a basis for shaping policy recommendations. The other main activity was the transfer of knowledge for energy advising to partners, teachers and students of vocational schools. The transfer was made through training for partners, who subsequently transferred the know-how to their local vocational schools through ten training events for teachers and students. After equipping students—trained as energy advisors—with the tools and techniques for visiting households, household visits commenced. Households were approached in cooperation with Centres for Social Work (an information flyer was prepared and the Centres collected households' applications for visits).

During the first visit, the advisors made an energy audit of the household and studied its habits. Based on these inputs, tailored advice was given to each household in order to empower them to reduce energy and water use. Apart from advice, the households also received free energy- and water-saving devices that helped them to make further savings.

## Types of Outcomes

Results for the overall project (for four countries combined) show that over 200 students and volunteers from vocational schools and faculties were trained to perform energy audits in energy-poor households. They helped partners to implement 1564 household visits, whereby basic energy efficiency measures were put in place and over 6650 free energy- and water-saving devices were installed. The investment of about €30 worth of free devices resulted in annual savings of over €65 in the visited households, or over €560 saved during the lifetime of devices. In total, €48,200 was invested in energy-saving devices that could save over €840,000 during the lifetime of the devices.

Overall, the initiative was considered successful. Apart from the practical quantitative level of reducing energy consumption in households, it was successful also on a structural level through informing policy issues. Participants found the visits very helpful, especially in terms of understanding their energy and water use better, but they also showed high appreciation for the free energy- and water-saving devices.

## CONCLUSION

Slovenia has a small energy sector, where oil with its 45% share represents the main energy source (mainly for transport). Electricity generation can be divided in three similarly sized parts: hydropower, nuclear energy and coal. Trends in energy policy, as prepared in the new proposal for the national energy concept, go in the direction of keeping the status quo regarding current energy sources, while also trying to follow EU directives and fulfil EU targets for renewable energy, energy efficiency and GHG emissions.

The emphasised SECI provides an example of initiative that is focusing on energy poverty and includes activities on practical and structural levels. The cross-national project's aim was to reduce energy consumption in low-income households, which it achieved through knowledge exchange and the provision of energy advice. Results show over €65 of savings per household per year on average, which is significant for many Slovenian households. It also included policy aspects, as advocacy activities were part of the project. Results from the practical part of the project were used for advocacy work and, in that way, decision-makers were presented with a 'ready-to-use' scheme. Engaging decision-makers on the national level and their activation on the topic of energy poverty, and subsequent preparation of a nationwide programme for energy poverty alleviation was one of the main successes of the REACH project.

## REFERENCES

ARSO. (2015). *Kazalci okolja: [EN30] Proizvodnja in raba električne energije (kazalec združuje kazalca EN12 in EN17)*. Available at http://kazalci.arso.gov.si/sl/content/proizvodnja-raba-elektricne-energije-kazalec-zdruzuje-kazalca-en12-en17?tid=4. Accessed 6 February 2019.

Export.gov (2017). *Slovenia—Electrical power systems and energy*. Available at https://www.export.gov/article?id=Slovenia-Electrical-Power-Systems. Accessed 6 February 2019.

SURS. (2018). *Količina energije, namenjene končni rabi, je v letu 2017 znašala 206.000 TJ.* Available at https://www.stat.si/StatWeb/News/Index/7722. Accessed 6 February 2019.

# CHAPTER 10

# From Efficiency to Sufficiency: Insights from the Swiss Energy Transition

*Laure Dobigny and Marlyne Sahakian*

**Abstract** In the wake of the Fukushima nuclear disaster, the Swiss 2050 Energy Strategy aims to promote energy efficiency, renewable energy (RE) sources, and nuclear power phase-out. Against that backdrop, this chapter provides a brief overview of the socio-material dynamics of household energy use in Switzerland, highlighting the role of regional energy providers, and the influence of building standards and social norms on everyday energy usage. We then examine current energy policies, before turning to the characteristics of Swiss sustainable energy initiatives—highlighting the important role of research—and focus on a best practice effort to support sufficiency-based transformations. We conclude with some reflections on the importance of focusing on sufficiency, rather than efficiency, in the framing and design of energy initiatives aimed at households.

**Keywords** Building standards · Energy transition · Energy policy · Action-research · Switzerland

L. Dobigny (✉) · M. Sahakian
University of Geneva, Geneva, Switzerland

M. Sahakian
e-mail: Marlyne.Sahakian@unige.ch

© The Author(s) 2019
F. Fahy et al. (eds.), *Energy Demand Challenges in Europe*,
https://doi.org/10.1007/978-3-030-20339-9_10

## INTRODUCTION

The Swiss energy system is relatively decentralized compared to neighbouring countries, with regional energy providers for electricity, gas, and district heating that generally take the form of private entities with the State as a primary stakeholder. Some private companies exist, as well as energy production cooperatives (citizen-led and producing renewables), but they are marginal compared to the role of public local companies in contributing to Swiss energy production and distribution. However, new consortiums between citizens and local public companies are emerging, particularly towards renewable energy (RE) generation and involving the implementation of RE plants with the financial participation of citizens. Through a citizen referendum in May 2017, Switzerland adopted the "2050 Energy Strategy" involving the following measures: lowering energy consumption, improving energy efficiency, and promoting RE. A progressive nuclear phase-out is also planned. Following a brief overview of socio-material dynamics of household energy use and energy policy in Switzerland, this chapter examines the characteristics of Swiss sustainable energy initiatives, and provides a good practice example of an initiative that seeks to contribute to sustainable transformation in the energy system through support for sufficiency-based lifestyles.

## SOCIO-MATERIAL DYNAMICS OF HOUSEHOLD ENERGY USE IN SWITZERLAND

A specificity in Switzerland is the system of central heating in buildings with estimated individualized heating bills for apartment units, and little to no thermostats. With the exception of metering in private homes, and the limited smart meters being deployed among households, there is little opportunity to apprehend detailed individual consumption for heating in apartments, as these are often calculated on an annual basis and bundled in with other utilities. Legislation in favour of individual bills for heating and hot water has been in place since 1993 for new building constructions. Further revisions of this legislation (the most recent one dated 2018) have included older building stock, albeit selectively. In buildings that boast the Swiss energy-efficiency label, Minergie, energy consumption is calculated by individual meters per apartment unit, and sophisticated floor heating systems result in the necessity to fine-tune hydraulic valves in order to adapt indoor temperatures.

The "performance gap" (Gram-Hanssen and Georg 2017) between these highly-energy-efficient buildings and usage is relatively well recognized by engineering companies and developers in Switzerland. Currently, heating is mainly provided by fossil fuels (primarily petroleum and gas) and electricity is mainly produced by hydroelectricity (60%) involving nuclear energy (32%) (SFOE 2018).

Beyond the material arrangements of energy systems and building standards, there are also less visible ways to shape energy demand: social norms and related prescriptions also have an influence on how energy-using practices play out. In Switzerland, social norms around cleanliness and tidiness appear to be quite strong, as the Swiss adage "*propre en ordre*" (clean and in order) expresses, which have implications for energy-related practices, including laundry and cleaning. Other normative ways of doing may have more positive implications for energy usage: it is not unusual for apartment buildings to have a shared laundry facility, for washing and drying clothes (e.g. in a heated room in the basement), although private ownership is on the rise. Laundry machine purchases have grown exponentially in the past decades, along with the normalization of dryers (Sahakian and Bertho 2018). Nevertheless, the new trend towards social and ecological building cooperatives means that collective laundry rooms are still being designed into these buildings, albeit on a small scale. In the mobility domain, while private cars are the preferred means of transport for many, public transport is quite efficient and car/bike sharing is becoming more popular. There are also public events to "slow down", involving biking and walking in city centres. In September 2018, and with the *Votation Vélo* (Bike Vote), 73.4% of Swiss voters agreed to integrate the right to bike lanes into the Constitution, on equal footing with pedestrian ways; yet how this will translate into local implementation remains to be seen.

## ENERGY POLICY IN SWITZERLAND

Partly due to a reaction to the Fukushima nuclear disaster, Switzerland has adopted a "2050 Energy Strategy" (Swiss Confederation 2016). In addition to reduced energy usage, mostly expected through efficient appliances and buildings, and the promotion of RE sources, new nuclear power plant construction is prohibited and a progressive nuclear phase-out is planned. To set up these objectives, the government is promoting refurbishment of buildings (by providing monetary incentives for owners to switch from oil heating systems to heat pumps, or towards insulation works); energy efficiency for appliances (through monetary incentives) as well as for cars (by enacting binding legislation); RE implementation (via the introduction of monetary schemes similar to feed-in tariffs in other European countries); the promotion of RE production and usage among households and at the neighbourhood level; and smart metering in households. Smart technologies are in a pilot phase, but a recent report by the Swiss government signals that such technologies are on the rise—put forward with the hope of engendering greater energy efficiencies in relation to systems of distribution. The Swiss government is also supporting research in energy transitions, focused mainly on technical innovations, but with more modest financial support dedicated to socio-economic aspects of energy production and consumption.

Due to the Swiss energy system, made up of mostly local and public utilities, and a participative democracy where citizens engage with energy issues (through regional and national referendums), there is a degree of trust in the local utilities by the general population, and towards the energy transition.

## TRENDS IN ENERGY INITIATIVES AIMED AT SWISS HOUSEHOLDS

In Swiss campaigns, and when it comes to tackling the demand side of energy usage, individual actions and efficiency measures are mainly being promoted: for example, turning off lights and appliances, changing old energy-intensive appliances for more efficient models (e.g. fridge), or other technical changes in the household (e.g. buying LED bulbs, by offering a special discount), among others. Local utilities, federal institutions, environmental NGOs and associations mostly lead these campaigns. There are also some efforts to promote solidarity between the so-called Global North and Global South: for example, initiatives to save energy in Switzerland, with savings invested in RE schemes elsewhere.

A key characteristic in Switzerland is the number of Sustainable Energy Consumption Initiatives (SECIs) that are led by academics and research teams. This is due to an ambitious national research

**Table 10.1**  Number of national SECIs in Switzerland according to their problem framing

| Problem framing | No. of initiatives |
| --- | --- |
| Changes in complex interactions | 12 |
| Changes in everyday life situations | 15 |
| Changes in individuals' behaviour | 9 |
| Changes in technology | 6 |

policy that supported research programmes on energy issues in the past decade, as mentioned above. As a result, numerous initiatives aimed at improving household energy usage (and systems more generally) are underway in Switzerland, engaging with innovative processes such living labs and action research, with new collaborations underway between municipal actors, energy companies and researchers (Sahakian and Dobigny 2019). A review of SECIs in Switzerland according to their problem framing reflects this trend: SECIs that promote changes in complex interactions and in everyday life situations (Table 10.1) are more numerous than in other European countries (Jensen et al. 2018).

Swiss SECIs also reflect the specificities of energy consumption in Switzerland, such as the significance of individual car usage. Several SECIs therefore propose initiatives aimed at changing mobility options, by promoting biking and e-biking, bike and car sharing, or public transport usage (for example, Bike4car and Publi Bike). Encouraging sufficiency measures or challenging the social norms tied up with un-sustainable energy usage are less common in the Swiss SECIs. Pumpipumpe offers a counter example, as detailed in the case study below.

## Case Study: Pumpipumpe[1]

The case study selected for Switzerland demonstrates how initiatives can focus on changes in everyday life situations, towards an overall aim of reducing (energy) consumption, but not based solely on information campaigns around energy or the introduction of more efficient appliances. Launched in Switzerland in 2012, Pumpipumpe is a platform to promote the sharing of appliances and other household items between neighbours. According to Sahakian (2017), the two founders were sharing a workspace in 2012 when they came up with the idea of creating a way for people to share everyday household items. In developing a sharing platform, the duo decided to work at the level of neighbourhoods, where they saw enormous potential. They came up with the idea of using the mailbox as a personal space and communication tool: the Pumpipumpe sticker system allows people to place specially designed labels directly on their mailbox, illustrating different household items (e.g. drill, ladder, books, and toys) that they are willing to share.

### Methods for Intervention

The aim of the project Pumpipumpe is to reduce the purchase of household items while promoting the sharing of consumer goods and community relations. A secondary objective is to prompt a change in how products are designed, with longevity-by-design in mind. The social dimensions of this initiative are put forward: sharing takes place through local networks, improving social interaction in urban neighbourhoods in geographic communities of place rather than solely online. An online platform was also launched allowing participants to order stickers, and more recently, in 2015, access an online interactive map, where people can see what items are available for sharing in their neighbourhood. Stickers can also be ordered in partner shops, from Pumpipumpe ambassadors, or from individuals who purchase several sheets of stickers to share with their contacts.

[1] http://www.pumpipumpe.ch.

### Framing the Energy Challenge

Pumpipumpe creates opportunities for sharing at the scale of a neighbourhood, so as to reduce the purchase of household items but also prompt people to think about product and service design, inspired by the cradle-to-cradle design philosophy. As quoted in Sahakian (2017), the co-founder states that "(s)haring makes sense in the material world as we have it today and would make sense in a world where we have very intelligent products, which go back into recycling". She feels that Pumpipumpe could ultimately influence purchasing decisions: "How we buy, how we select. If you see that there are already two pasta machines in your neighbourhood, you know you don't need to buy one", leading to less material throughput in the economy and less waste.

The initiative has been designed to be simple, accessible to all people, and easy to use. That is why the stickers are composed solely of graphically-designed images (without any text) to facilitate the exchange between neighbours, without difficulties of language or legibility (e.g. easily understandable by children or old people). Participants are free to share their tools, kitchen appliances or toys with their neighbours, however they see fit. This way, Pumpipumpe promotes the free sharing (not renting for money) of personal belongings.

### Outcomes and Outputs

More than 15,000 households in Switzerland, Germany, and Austria have ordered Pumpipumpe stickers, and approximately 9450 addresses are listed on the online platform. However, the association does not keep track of who is using the platform and what sharing activities are actually taking place. In the future, the co-founder envisions a smartphone application for identifying different available items in a given area, but for this, additional funding would be necessary. Beyond virtual connections, she emphasizes the importance of the "real, live network" as she put it:

> We live in rather anonymous urban neighbourhoods. We have a beautiful digital network, all over the world [...] With social networks, we are connected with people like us everywhere in the world, but in fact in the real network, in the neighbourhood, it is the opposite. Because it is very

diversified: there are people with different political points of view, different ages… quite different precisely. But in the same place. So that's exactly the opposite of digital social networks, and […] I think it's important to connect with people who are different, who do not have the same opinions about everything, who do not have the same habits. And not to be afraid of this discussion, of this interaction with quite different people.[2]

Thus, without large costs, the initiative aims to have a high impact on practices and representations of ownership, consumption, and sharing, and promotes a sufficiency-based lifestyle. There is also coherence between the message (less consuming and more sharing) and the action type (simple to use, appropriable by all, and exchange facilitating). This type of initiative can have an impact on a regional or international scale, as citizens across Europe and elsewhere have already demonstrated their interest in ordering Pumpipumpe stickers. This international diffusion and success is no doubt due to the simplicity of its design and functioning.

## CONCLUSION

Challenges to achieving the Swiss energy transition and objectives of "2050 Energy Strategy" are particularly related to space heating, mobility, and RE roll out. The specificity of heating in apartment buildings poses a particular challenge: the lack of a resident's control over their own heating system, and possibility to set indoor temperatures, implies that people would have a difficult time reducing indoor temperatures if they wish to do so. Empowering people to be able to adapt their indoor temperatures is a necessary step towards engaging households in the energy transition (Dobigny 2016, 2017). The Swiss energy system, mostly based on local and public utilities, also has an important role to play in the energy transition, particularly due to the trust of population in these local utilities. An interesting development has emerged in relation to renewable energies: to achieve the "turn" towards renewable energy resources, collaborations are underway between utility providers and citizen groups. This is an original development in Switzerland, situated between large-scale RE implementation led by utility companies, and small-scale citizen cooperatives.

[2] Interview extract, translated from French. Zurich, Switzerland, December 15, 2017.

The utility companies are also actively engaged in campaigns, initiatives, and R&D to decrease energy usage, often collaborating with community associations or researchers towards this aim. A particularity of the Swiss case is indeed the important role of academics and research teams when it comes to implementing and designing energy initiatives, or re-thinking our relationship to energy and absolute reductions in innovative ways. This is due to an ambitious national research policy that supported research programmes on energy issues in the past decade, with a specific focus on socio-economic approaches to energy, in addition to technical developments. This lesson from Switzerland could inform European policy towards further supporting research-action initiatives drawing from social sciences and the humanities. Another lesson learned from the Swiss SECIs to inform European policy is to account for coherence: much like the Pumpipumpe example given above, the design and functioning of any such initiatives should be coherent. If less energy usage is an aim, more initiatives need to focus on sufficiency, rather than efficiency—and should be designed accordingly.

## REFERENCES

Dobigny, L. (2016). *Quand l'énergie change de mains. Socio-anthropologie de l'autonomie énergétique locale au moyen d'énergies renouvelables en Allemagne, Autriche et France* (Thèse de Sociologie). Université Paris 1 Panthéon-Sorbonne.

Dobigny, L. (2017). *What renewable energy change? The role of technical systems in energy uses and representations.* 13th Conference of the European Sociological Association (ESA), Athens, Greece, 29 August–1 September.

Gram-Hanssen, K., & Georg, S. (2017). Energy performance gaps: Promises, people, practices. *Building Research & Information, 46*(1), 1–9.

Jensen, C. L., Goggins, G., Fahy, F., Grealis, E., Vadovics, E., Genus, A., et al. (2018). Towards a practice-theoretical classification of sustainable energy consumption initiatives: Insights from social scientific energy research in 30 European countries. *Energy Research & Social Science, 45*, 297–306.

Sahakian, M. (2017). Toward a more solidaristic sharing economy: Examples from Switzerland. In M. J. Cohen, H. S. Brown, & P. J. Vergragt (Eds.), *Social change and the coming of post-consumer society: Theoretical advances and policy implications.* Oxon and New York: Routledge.

Sahakian, M., & Bertho, B. (2018). *L'électricité au quotidien: le rôle des normes sociales pour la transition énergétique suisse.* Genève, Suisse, Fonds national suisse (FNS), PNR71. https://www.unige.ch/sciences-societe/socio/files/9415/3502/7352/Brochure_PNR71_DEF.pdf.

Sahakian, M., & Dobigny, L. (2019). From governing behaviour to transformative change: A typology of household energy initiatives in Switzerland. *Energy Policy, 129,* 1261–1270.

Swiss Confederation. (2016). Energy law, (LEne) of 30th septembre 2016 (entered in force on 1 January 2018), Switzerland. https://www.admin.ch/opc/fr/classified-compilation/20121295/index.html.

Swiss Federal Office of Energy (SFOE). (2018). *Statistique suisse de l'électricité 2017.* http://www.bfe.admin.ch/index.html?lang=en.

Vuille F., Favrat, D., & Erkman, S. (2015). *Comprendre la transition énergétique.* Presses polytechniques et universitaires romandes.

# CHAPTER 11

# Sustainable Energy Consumption and Energy Poverty: Challenges and Trends in Bulgaria

*Marko Hajdinjak and Desislava Asenova*

**Abstract** The chapter provides a short overview of the sustainable energy consumption challenges in the Bulgarian residential sector, with a special focus on the issue of energy poverty. The chapter first looks at the main characteristics of the household energy consumption (energy mix, use of renewables, socio-material factors) and then summarises the relevant information about the Bulgarian energy system and energy policies. The authors discuss the most important findings from the review of sustainable energy consumption initiatives (SECIs) involving Bulgarian households and present a good practice example of one such initiative (European Citizens Climate Cup [ECCC]). The conclusion of the chapter considers why Bulgarian households rarely take measures aimed at increasing their energy efficiency.

M. Hajdinjak (✉) · D. Asenova
Applied Research and Communications Fund, Sofia, Bulgaria
e-mail: marko.hajdinjak@online.bg

D. Asenova
e-mail: desislava.asenova@online.bg

115

**Keywords** Energy poverty · Renewable energy sources · Inefficiency · Transition and reforms · Barriers and challenges

## INTRODUCTION

This chapter presents a short overview of the sustainable energy consumption challenges in Bulgaria. It includes a brief description of the national energy system and energy policies, brings forth some findings from the review of sustainable energy consumption initiatives (SECIs) involving Bulgarian households, and presents a good practice example of one such SECI.

## SOCIO-MATERIAL DYNAMICS OF HOUSEHOLD ENERGY USE IN BULGARIA

The residential sector is the third largest sector in terms of final energy consumption (24%), after transport (35%) and industry (28%), and ahead of services (11%), and agriculture (2%) (Odyssee-Mure 2015). A relatively high share of renewable energy sources (RES) in the energy mix of households (Table 11.1) is explained by the fact that the most commonly used fuel for heating of homes is wood (59%). This is not valid for the capital Sofia, where 60% of households use district heating, but in rural areas almost all dwellings use either firewood (63%) or coal (32%) as the main heating source (National Statistical Institute of Bulgaria 2011; Eurostat 2016).

As Table 11.1 shows, renewable energy is the second most important fuel source in the residential sector. This has helped Bulgaria to already exceed its 2020 target of at least 16% energy consumption coming from RES (RES provided 19% of the final energy consumption in 2015).

**Table 11.1** Final energy consumption in households by fuel (2016)

| Fuel | Bulgaria (%) | EU-28 average (%) |
|---|---|---|
| Electrical energy | 42 | 24 |
| Renewable energy | 33 | 16 |
| Derived heat | 15 | 8 |
| Solid fuels | 6 | 3 |
| Gas | 2 | 37 |
| Total petroleum products | 2 | 12 |

*Source* Eurostat (2018)

However, wood, wood wastes, and vegetable wastes account for almost 80% of the renewables balance sheet for 2016. Most of this biomass is consumed in the residential sector for heating—especially in rural areas and small towns, where many houses are heated by old and ineffective stoves (Gantcheva 2018). The widespread use of cheap and low quality wood has had a negative effect on the air quality in recent years. Another problem is that in addition to the state-controlled harvesting, distribution, and sale of wood (regulated by the Ministry of Agriculture and Food and the State Forestry Agency), up to 25% of firewood is harvested illegally[1] (Boonstra et al. 2015).

The Bulgarian housing stock has in general very poor energy efficiency performance. 65% of the 3.9 million housing units were built before 1990, including over 800,000 households located in prefabricated multi-storey buildings in poor condition and with inefficient or non-existing thermal insulation (Gaydarova 2012).

Energy poverty is a significant problem. According to EU Energy Poverty Observatory data for 2016, 41% of Bulgarians cannot maintain adequate thermal comfort in their households.[2] The rising electricity and district heating prices in recent years have forced many households towards using coal and wood for heating, which further worsens air and living quality. Although electricity prices are still the lowest in the EU (less than half of the EU average of 0.20 Euro per kWh), incomes are also significantly below the EU28 average, which means that energy costs represent a considerable burden on family budget. Over 400,000 households are claimed to be highly vulnerable to increases in electricity prices, while another 149,000 households are income-poor (Export.gov 2017).

Vulnerable consumers are often prepared to compromise their energy comfort and expose themselves to health risks in order to cut their energy expenses. A widespread practice of underheating to reduce energy bills has been observed. Therefore, special attention needs to be taken to ensure that households do not curtail their energy use in a way that would jeopardise their health or well-being.

Environmental concerns are in recent years becoming an important motivator to save energy, but the main incentive for a typical Bulgarian household is to cut expenses. Nevertheless, many households remain

---

[1] This explains the poor quality of the wood on the market (wet, young and poorly chopped) contributing to the high pollution when burning.

[2] See https://www.energypoverty.eu/indicator?primaryId=1461.

inactive, believing that energy saving can only be achieved after a large initial investment (e.g. purchasing energy efficient appliances, retrofitting of homes), which is often unaffordable. A typical Bulgarian household can therefore save energy only through the application of measures that can be performed with little or no cost. In addition, the artificially low electricity prices are a negative incentive for changing the behaviour even in households with moderate to high incomes.

One of the ways the government is addressing energy poverty is by subsidising the prices of electricity. This measure, however, exacerbates a bad situation for the state-owned energy holding company (Bulgarian Energy Holding [BEH]), which is at a permanent financial loss. It also sends negative signals to foreign investors, who perceive their potential involvement in the Bulgarian energy system as risky and unattractive (Boonstra et al. 2015).

The 2013 protests over the rise of energy prices that ended with the resignation of the government and early elections are a constant reminder to the authorities that a transition from regulated, centralised, supply-based energy system, towards liberalised, decentralised, and prosumer-focused energy can lead to social instability if not handled properly (Vladimirov et al. 2018).

## ENERGY POLICY IN BULGARIA

According to the last version of the Bulgaria's Energy Strategy, the main efforts for developing the energy sector are directed towards energy efficiency, energy self-sufficient buildings, electric vehicles, renewable energy, and building of smart grids (MEET 2011). In order to comply with the EU Directive 2012/27/EU that aims to establish a common framework to promote energy efficiency within the EU, Bulgaria has developed the National Energy Efficiency Action Plan 2014–2020 (Ministry of Energy 2017). The NEEAP defined the long-term strategy, main actors, objectives, and measures for four main energy consuming sectors. In the residential sector (with a strong focus on multifamily buildings), minimum energy performance standards have been defined, and necessary economic incentives and financing instruments established. For example, regarding domestic appliances, eco-design requirements and energy labelling were introduced (Energy Efficiency Watch 2013).

A recent assessment of energy policy in Bulgaria highlights the steady progress of the country regarding the greening of its energy and economy, being one of the first EU members to meet its 2020 targets for RES (Center for the Study of Democracy 2017a). The report highlights four main long-term energy risk factors:

- Energy poverty: 41% of households are not able to keep their homes adequately warm and 29% have arrears on utility bills;
- Energy intensity of the economy: it remains above the EU average despite continuous improvements;
- Low level of diversification: especially in the natural gas sector, which depends on Russia as the single source of gas supplied through a single route; and
- Bad governance: corruption, bad policy choices and incompetency have considerably contributed to the recent energy price increases in the country.

The largest player in the electricity market is Bulgarian Energy Holding (BEH), which owns a diverse group of companies engaged in electricity generation, supply and transmission. Electricity is mainly generated by coal burning power plants (43% in 2017) and nuclear power (36%). Hydropower supplies 8% of electricity, while additional 8% is generated by wind, solar power and biomass (Global Legal Insights 2019).

District heating networks exist in 12 Bulgarian cities, serving in total about 600,000 households. *Toplofikatsia* Sofia, which provides district heating in the capital, is 100% owned by the Municipality of Sofia, but the central heating companies in Plovdiv and Varna (second and third largest cities) are privately owned. Regardless of the ownership, all district heating companies are local monopolies. Most use natural gas as fuel, although a few still use coal.

The rise in renewable energy in the overall mix has been mainly driven by an increase in the use of solid biomass, with a rise in wind and solar PV capacity also noticeable. Electricity from RES has been supported since 2007 through a preferential feed-in tariff scheme, the obligation for energy distributors to connect green energy producers to the grid, and the creation of long-term loan guarantees for banks financing wind and solar power plants (Boonstra et al. 2015).

These measures have resulted in a surge of wind and solar plants being constructed (almost 90% of all RES generation capacity was installed in 2010–2012), but the majority of them are owned by a handful of suppliers, who often benefit from access to favourable processing and procedural conditions. As a result, these developments did not contribute to decentralisation of electricity production. Another consequence of this faulty process was a sharp increase in final user tariffs, which coincided with the peak of the economic recession, leading to a wide popular backlash against green energy. The feed-in tariff scheme was finally suspended in 2015, but is yet to be replaced with a new state support scheme. This contributed to the minimal uptake of RES sources since 2015 (Vladimirov et al. 2018).

Furthermore, administrative procedures for installation and exploitation of small PV capacities are among the most discouraging and burdensome in the EU, and not surprisingly, investments into new capacities have been in a rapid decline (Vladimirov et al. 2018).

## TRENDS IN HOUSEHOLD ENERGY CAMPAIGNS IN BULGARIA

National energy campaigns are mainly focused on cutting down household energy consumption, reducing greenhouse gas emissions, and promoting green transport. They are organised by a range of actors including government, municipalities, NGOs, local communities, and businesses. Many are implemented as part of EU funded projects.

The largest initiative in terms of invested funds and involved households is the National Programme for Energy Efficiency of Residential Buildings, which provides grants for renovation of multifamily residential buildings, thus improving their energy efficiency. The programme targets over 2000 buildings in Bulgaria, which equals over 100,000 households (Ministry of Regional Development and Public Works 2015). The Energy Efficiency and Renewable Sources Fund (EERSF) was established in 2005 by the Bulgarian government to provide funding and technical assistance for energy efficiency projects implemented by public (municipalities, universities, hospitals) and private sector (businesses and private households).[3]

---

[3] See https://www.bgeef.com/en/.

Non-governmental organisations are often an important driving force in energy campaigns. For example, the environmental Association *Za Zemiata* (Friends of the Earth—Bulgaria) focuses on developing strategies to involve citizens in energy management, and brings together representatives of civil society, local authorities, business, and scientists to help build local energy cooperatives.[4] One of the rare examples of larger-scale cooperation on the local level is the Municipal Energy Efficiency Network (EcoEnergy). Registered in 2003, the network brings together Bulgarian municipalities to promote the efficient use of fossil fuels, increase the use of renewable energy and improve the energy security of municipalities. EcoEnergy is a supporting structure of the Covenant of Mayors and implements a number of energy related projects funded by European programmes.[5]

Nonetheless, community based sustainable energy initiatives in Bulgaria are often hindered by the inert political and bureaucratic system, numerous administrative barriers and legal hurdles, high investment costs, and unpredictability of the energy sector.

CASE STUDY: EUROPEAN CITIZENS CLIMATE CUP (ECCC)[6]

European Citizens Climate Cup (ECCC) was a competition of private households within and between countries with the target to achieve the highest energy savings. The competition attracted 8400 households from 11 European countries and regions. It was financially supported by the Intelligent Energy Europe programme. The competition lasted from April 2011 to April 2012.

A variety of incentives motivated households to participate. These included an appeal to patriotic feelings (being part of a national team and compete against other countries), a possibility to win attractive awards, a financial incentive (saving energy in order to save money), and a chance to contribute to a better and healthier environment. Following a disappointing response to initial recruitment efforts, Sofia Energy Agency (SOFENA), the Bulgarian partner in ECCC, enlisted the help of municipalities and corporate businesses to recruit households. In the end, 1006 Bulgarian households participated in the initiative, easily exceeding the target of 750.

---

[4] See https://www.zazemiata.org/.

[5] See http://www.ecoenergy-bg.net/en.

[6] See  https://ec.europa.eu/energy/intelligent/projects/en/projects/eccc  for  more information.

All participating households had to open an Energy Saving Account (ESA). ESA stored and analysed their energy consumption and cost data, and calculated $CO_2$ emissions using national $CO_2$ indexes and climate factors. This information was used to inform each household about its environmental impact and to propose concrete actions to lower electricity and heat energy consumption. ECCC gave users tailor-made advice for modifying the ways they cook, wash, iron and consume water.

The ECCC campaign actively cooperated with media (print, online, and TV) and was endorsed by numerous corporate actors, municipalities, schools, citizens' associations and other multipliers. The Bulgarian ECCC website and social profiles were updated frequently with energy saving tips, success stories and testimonials from participants. Regular energy events were organised, including lotteries with sponsored gifts. The success of the campaign was monitored by brief weekly and more detailed monthly reports displaying benchmarks in energy consumption according to different user groups (tenants/owners), different building types, or different energy sources. The competition ended in April 2012, when the winners were announced and awarded.

### Framing the Energy Challenge

As households account for almost 30% of the $CO_2$ emissions in the EU, the central aim of the initiative was to make people aware of their personal impact on the climate change and motivate them to implement energy improvements in their own situation. Apart from trying to influence and change the individual energy-using actions, the initiative also had a strong community aspect, as participants were encouraged to compete with other Bulgarian participants, but also as a national team competing with other countries. Householders actively used online tools to record consumption data, they received feedback on trends, and participated in social media, engaging in discussions and sharing their experiences with energy savings measures. They also attended different events including lectures, meetings, fairs, and workshops. At the end of the competition, householders completed an evaluation survey.

The campaign was predominantly successful. Bulgaria achieved better than planned results in electricity savings (5.81% against 2% targeted), however, savings of heating energy at 0.52% were considerably lower than the expected 4%. Partial explanation for this is that electricity consumption per household in Bulgaria is higher than in any other ECCC country. Consequently, there were untapped reserves for electricity savings, which the competition managed to activate. On the other hand, opportunities for heating savings were much more restricted, due to the unfavourable state of Bulgarian building stock and limited financial ability of residents to invest in retrofitting. Hence, heat energy could often be saved only at the expense of thermal comfort (Julius 2012).

The winner of the Bulgarian ECCC was a family from a small town in Western Bulgaria, living in a detached two-storey building constructed in the 1980s with no energy efficiency measures. During the campaign, the family invested 1665 EUR into retrofitting their home (efficient lighting, new appliances like refrigerator, cooker and washing machine, and a solar panel on the roof). These measures decreased their energy consumption by 54%, with additional 14% achieved through behavioural changes.

## Conclusion

Of 45 SECIs in Bulgaria, examined and described in the frame of the ENERGISE project (Table 11.2), 14 have been classified as 'Changes in Technology,' 19 as 'Changes in Individual Behaviour,' 6 as 'Changes in Everyday Life Situations' and 6 as 'Changes in Complex Interactions.' The objectives of the majority of initiatives implemented in Bulgaria are therefore to influence attitudes and choices related to energy efficiency and potentially change the energy consuming behaviour of individual households or household members, or to achieve energy savings through introduction of energy efficient technical measures. Only a minority of initiatives target more complex solutions that necessitate active involvement of a community of people who do not necessarily know each other and are willing to act and interact for the common good, and not only to reduce the energy costs of their own household.

Another curious feature of Bulgarian SECIs is that most of them (32) are implemented as part of international projects—mostly EU funded. Only a few initiatives are true grass-root projects developed and implemented by the household residents themselves. An interesting observation from the analysis on international projects is that Bulgarian householders are often very eager and active participants in top-down initiatives (in many projects, especially the ones involving a competitive dimension of energy saving, Bulgaria archived higher than average levels of participation and some of the best results), but are very reserved when it comes to self-organisation and cooperation with their neighbours and co-citizens.

**Table 11.2**  Number of national SECIs in Bulgaria according to their problem framing[a]

| Problem framing | No. of initiatives |
| --- | --- |
| 🔄 Changes in technology | 14 |
| 🔴 Changes in individuals' behaviour | 19 |
| 🟡 Changes in complex interactions | 6 |
| 🟢 Changes in everyday life situations | 6 |

[a]See http://www.energise-project.eu/projects for explanation

An additional reason why Bulgarian households rarely take measures aimed at increasing energy efficiency is the widespread perception that ordinary citizens cannot change anything, as the energy sector is completely controlled by the state and energy monopolies. Substantial legislative barriers and regulatory burdens further discourage Bulgarian households from taking action—a case in point is the SECI 'Solar Roof.' Households from a 15-storey apartment building in Sofia jointly installed 120 solar panels on the roof of the building. While the purchasing and installing the panels took two weeks, obtaining a considerable number of different permits took almost two years.

Finally, there is the crucial issue of low incomes and widespread (risk of) energy poverty. The main priority for most households is therefore not cleaner energy and protection of environment, but lower energy expenses. It is not surprising that a considerable number of initiatives aim to reduce the energy costs of households, mostly through measures like retrofitting and thermal insulation of multi-storey residential buildings (typically through grants provided by the Bulgarian state and EU funds). The strong focus on technological solutions and retrofitting is also a consequence of the old age and poor state of repair of the building stock, which mostly dates from the socialist period, when energy was cheap and plentiful.

# REFERENCES

Boonstra, B., et al. (2015). *Social innovation in energy supply from a European and global perspective* (Restricted state-of-the-art report of project SI-DRIVE. Social Innovation: Driving Force of Social Change).

Center for the Study of Democracy. (2017a). *A roadmap for the development of the Bulgarian electricity sector within the EU until 2050: Focus on fundamentals* (CSD Policy Brief No. 70). Available at https://csd.bg/fileadmin/user_upload/publications_library/files/23263.pdf.

Center for the Study of Democracy. (2017b). *Decentralisation and democratisation of the Bulgarian electricity sector: Bringing the country closer to the EU climate and energy core* (CSD Policy Brief No. 79). Available at https://csd.bg/fileadmin/user_upload/publications_library/files/2018_07/BRIEF_79_ENG.pdf.

Energy Efficiency Watch. (2013). *Energy efficiency in Europe. Assessment of energy efficiency action plans and policies in EU member states 2013.* Country Report Bulgaria. Available at http://www.energy-efficiency-watch.org/fileadmin/eew_documents/Documents/EEW2/Bulgaria.pdf.

Eurostat. (2016). *Share of fuels in the final energy consumption in the residential sector for space heating.* Available at https://ec.europa.eu/eurostat/statistics-explained/index.php?title=File:Share_of_fuels_in_the_final_energy_consumption_in_the_residential_sector_for_space_heating,_2016_(%25).png.

Eurostat. (2018). *Final energy consumption in households by fuel.* Available at https://ec.europa.eu/eurostat/tgm/table.do?tab=table&plugin=1&language=en&pcode=t2020_rk210.

Export.gov. (2017). *Bulgaria—Power generation.* Available at https://www.export.gov/article?id=Bulgaria-Power-Generation-Oil-and-Gas-Renewable-Sources-of-Energy-and-Energy-Efficiency.

Gantcheva, N. (2018). *Enable.EU project. D5.2. Case study report on governance barriers to energy transition.* Country report for Bulgaria. Available at http://www.enable-eu.com/wp-content/uploads/2018/10/ENABLE.EU-D5.2.zip.

Gaydarova, E. (2012). *Energy saving measures in residential buildings in Bulgaria.* Available at http://bpie.eu/wp-content/uploads/2015/10/E-Gaydarova_Bulgaria.pdf.

Global Legal Insights. (2019). *Energy 2019.* Bulgaria. Available at: https://www.globallegalinsights.com/practice-areas/energy-laws-and-regulations/bulgaria#chaptercontent1.

Julius, C. (2012). *European Citizens Climate Cup Final Report: 8,400 European citizens participated and saved energy!* Available at https://ec.europa.eu/energy/intelligent/projects/sites/iee-projects/files/projects/documents/eccc_8400_european_citizens_saved_energy_en.pdf.

Ministry of Economy, Energy and Tourism. (2011). *Energy strategy of the Republic of Bulgaria till 2020: For reliable, efficient and cleaner energy.* Available at https://www.me.government.bg/files/useruploads/files/epsp/23_energy_strategy2020%D0%95ng_.pdf.

Ministry of Energy. (2017). *National Energy Efficiency Action Plan 2014–2020.* Updated 2017. Available at https://www.me.government.bg/files/useruploads/files/npdee_2017_en.pdf.

Ministry of Regional Development and Public Works. (2015). *Energy efficiency of multi-family residential buildings national programme.* Available at http://www.mrrb.government.bg/en/energy-efficiency/energy-efficiency-of-multi-family-residential-buildings-national-programme/.

National Statistical Institute of Bulgaria. (2011). *2011 population census in the Republic of Bulgaria (Final Data).* Available at http://www.nsi.bg/census2011/PDOCS2/Census2011final_en.pdf.

Odyssee-Mure. (2015). *Average energy consumption per dwelling.* Available at http://www.odyssee-mure.eu/publications/efficiency-by-sector/households/average-energy-consumption-dwelling.html.

Vladimirov, M., et al. (2018). *Development of small-scale renewable energy sources in Bulgaria: Legislative and administrative challenges.* Sofia: Center for the Study of Democracy. Available at https://csd.bg/fileadmin/user_upload/publications_library/files/2018_07/DECENTRALISATION_ENG.pdf.

CHAPTER 12

# Finnish Energy Policy in Transition

*Eva Heiskanen, Senja Laakso*
*and Kaisa Matschoss*

**Abstract** In Finland, energy policy is in transition towards integrating energy projects in broader sustainability, liveability and innovation contexts. While energy saving has been pursued for decades, it is now part of a broader tendency in urban planning to promote sustainable lifestyles. Transition manifests in local actors' redistribution of power, challenging conventional ways of infrastructure development, forging new networks, and seeking novel solutions. The experimental case presented in the chapter, Smart Kalasatama, shows that local governments are close to citizens and, therefore, can influence the conditions for sustainable consumption and quality of life. Although they have an important role in energy policy, they still might lack the resources, expertise and the power to innovate, to evaluate projects, and in particular, to scale up innovative practices.

E. Heiskanen (✉) · S. Laakso · K. Matschoss
Centre for Consumer Society Research,
University of Helsinki, Helsinki, Finland
e-mail: eva.heiskanen@helsinki.fi

S. Laakso
e-mail: senja.laakso@helsinki.fi

K. Matschoss
e-mail: kaisa.matschoss@helsinki.fi

© The Author(s) 2019                                           127
F. Fahy et al. (eds.), *Energy Demand Challenges in Europe*,
https://doi.org/10.1007/978-3-030-20339-9_12

**Keywords**  Energy conservation · Energy policy · Experimentation · Transition · Urban climate action

## INTRODUCTION

Finnish energy policy is undergoing a period of transition. Previously, policy focused on the needs of industry, which consumes almost half of all the energy used in the country (Statistics Finland 2018a). The current government aims to increase renewable energy production to more than 50%, phase out coal and halve the use of mineral oil (MoEE 2017). These developments place new challenges on energy policy, where among other issues, energy provision in urban areas has re-emerged as an issue, after more than half a century of stability. For example, the envisaged coal phase-out problematizes the district heating system of cities like Helsinki, where district heating is still largely produced with coal-fired combined heat and power (CHP) combustion.

Home heating and domestic electricity use have been subjects of energy efficiency policy, but not at the top of the energy policy agenda until recently. A recent development is the increasing engagement of cities and rural municipalities in climate policy, featuring several nation-wide programmes in which cities and municipalities have committed to climate targets and engaged in a joint search for new solutions to decarbonize the built environment (Mickwitz et al. 2011).

This penultimate chapter highlights how Finnish SECIs reflect recent developments in Finnish energy policy. The most recent SECIs are largely locally based, combine energy saving with other concerns, and aim to develop combinations of technical and social solutions from the bottom up.

## SOCIO-MATERIAL DYNAMICS OF HOUSEHOLD ENERGY USE IN FINLAND

Finnish residential buildings are relatively energy efficient, because about 75% of the building area was constructed after the 1970s (Statistics Finland 2018a), when energy efficiency requirements were tightened. Owing to the high level of insulation and the wide diffusion of district heating, Finns are accustomed to stable indoor environments and well-functioning, automatized systems. Like other Nordic countries, average indoor temperatures are rather high (about 21 °C) in Finland (Karjalainen 2009).

However, there are two distinct cultures of home heating in Finland. Finnish apartment buildings are mainly served by district heating. These buildings (both owner-occupied and rented) are collectively managed and billed for district heating as one unit. Residents do not pay individually for their heating—rather, billing for space heating is by square meters (Matschoss et al. 2013). Apartments are equipped with thermostats, making the heating to some extent adjustable, but residents rarely adjust their heating systems (Karjalainen 2009). Until now, district heat has been relatively cheap in large cities due to the widespread use of CHP. Because of this, city dwellers in particular are not too concerned about energy costs.

Finnish detached houses, making up 55% of the residential building area (Statistics Finland 2019), are not usually served by district heating. Direct electric heating is still the most common heating source, often coupled with fireplaces. However, heat pumps have rapidly gained ground. About 900,000 heat pumps have been sold in Finland, providing about 15% of residential heat consumption (SULPU 2019), reflecting the Finns' propensity to rapidly adopt technological novelties. Energy costs are a much larger concern in detached houses than in apartments, especially for residents with electric or oil heating. Indeed, while energy poverty is relatively rare in Finland, rural elderly people with outdated heating systems are vulnerable, since it is difficult to afford major heating system investments in declining rural areas (Runsten et al. 2015).

Saunas are a distinct Finnish cultural peculiarity. Nevertheless, compared to space heating and domestic hot water, saunas are not a major consumer of energy (Statistics Finland 2018a), even though there are 2 million of them. However, since most modern saunas in cities are powered with electricity, they contribute to peak electricity (Järventausta et al. 2015). Individual saunas started to become a standard feature also in apartments, though this trend is declining in cities due to space constraints. In Helsinki, public saunas have made a comeback, reflecting a wider trend of new urbanism, which emphasizes shared services and amenities, including traditional bricks-and-mortar based, as well as new digital solutions.

ENERGY POLICY IN FINLAND

Finnish energy policy is in transition. For decades, policy focused on the needs of industry, which consumes more than 40% of all the energy used in the country (Statistics Finland 2018b). The share of renewable energy has grown steadily since the late 1970s, but much of it still comes from forest residues used by the pulp and paper industry. However, policymakers have gradually grasped that other renewable energy sources need to be developed, and increasing support has been directed to the development of wind power. Renewable energy amounted to 36% of the total energy production in 2017 (Statistics Finland 2018c). The current government aims to increase renewable energy sources to more than 50% and increase domestic energy provision to more than 55% by 2030. Additionally, Finland aims to phase out coal and halve the use of mineral oil (MoEE 2017). These developments place new challenges on energy policy. Among others, the envisaged coal phase-out challenges the district heating system of cities like Helsinki, where a large share of district heating is produced with coal-fired CHP combustion. This has been a cheap and reliable source of energy for decades, and when introduced, it cleaned up the air in cities (Apajalahti 2018). However, bioenergy based CHP is not deemed feasible for Helsinki—hence, the capital city is plunged into a search for new heating solutions: ideally ones that combine the flexibility and amenity of district heating with fossil-free energy sources like heat pumps (Rinne et al. 2018).

The dominant approach to energy has emphasized technological advances. Indeed, energy efficiency and renewable energy have gained momentum from the notion that Finnish export industries might benefit from innovation. There is growing consideration for behavioural, situational and systemic approaches, but none of these has yet systematically permeated the energy policy mindset (Karhunmaa 2018; Haukkala 2018).

Local authorities have an important, but hitherto somewhat neglected role in energy policy. Municipalities own a large share of the energy production and distribution system. Moreover, municipalities have an important role in delivering energy efficiency policies through land use planning, detailed city plans, and implementation of the requirements set out in the building code. Moreover, large cities develop land for construction and allocate it to construction companies, and can place requirements on new buildings through land release requirements. For example, they can place their own more stringent requirements concerning energy efficiency or building systems that enable residents to control their energy use.

Energy companies have a responsibility to provide energy advice to their customers. Virtually all Finnish electricity consumers have automatic meter reading installed, so most households can view their historical and comparative electricity consumption data online in almost real time. There are also several developments ongoing in developing smart district heating systems. Demand response (flexible use of heat and power depending on supply and demand) has become a hot topic in quite recent years, since it is considered necessary to prepare for a large increase in intermittent power production (Annala et al. 2018).

Citizens are highlighted in policy rhetoric (e.g. NEEAP 2017), but citizen movements focusing explicitly on energy efficiency have only recently emerged, since energy has been seen more as an expert domain. Residents' associations have shown some activism around particular solutions and technologies (Heiskanen et al. 2011), often gaining momentum from municipal-level climate action (Heiskanen et al. 2015). Another example of recent citizen activism are online discussion forums, for example a vibrant and popular discussion forum on heat pumps (Hyysalo et al. 2013).

## TRENDS IN NATIONAL HOUSEHOLD ENERGY CAMPAIGNS IN FINLAND

Energy saving was heavily emphasized during the oil crises, with forceful advice campaigns and restrictions on e.g. indoor temperatures. As oil prices and overall oil dependency declined, the tone of national energy campaigns shifted to technological innovation and energy efficiency, emphasizing that energy can be saved without loss of comfort.

Finnish national energy campaigns are mainly organized by Motiva, a state-owned company promoting energy efficiency, renewables and materials efficiency. Campaigns have not been a strong focus in recent years, but rather the provision of local targeted practical advice and engagement. This advice focuses on sensible use of energy, i.e. auditing, metering, training, automation, adjusting controls, refurbishment and renewable energy—and, most recently, demand response. *Energy Saving Week* is one of the nationwide campaigns for homes and workplaces. In addition to Motiva, energy companies also organize campaigns, such as the *Energy Family* competition by Vattenfall. The Finnish Environment Institute, coordinator of a large carbon neutral municipalities programme, has also organized various campaigns such as joint purchasing of solar panels.

Older SECIs are more focused on technology or individual behaviour change, whereas newer ones focus more on everyday practices and complex interactions between households and systems of provision (Table 12.1). There is a development towards living lab types of

**Table 12.1**  Number of national SECIs according to their problem framing

| Problem framing | No. of initiatives |
|---|---|
| Changes in complex interactions | 9 |
| Changes in everyday life situations | 13 |
| Changes in individuals' behaviour | 15 |
| Changes in technology | 10 |

approaches, i.e. testing technologies in real-life contexts by engaging citizens in experimentation towards more sustainable lifestyles (Laakso et al. 2017). Moreover, there is a tendency towards integrating energy projects in broader sustainability, liveability and innovation contexts. Many of the newer SECIs still focus on technology, but with the engagement of users, their everyday practices and sometimes even addressing the complex interactions between technologies, hence they are categorized as 'changes in complex interactions' and 'changes in everyday life situations'.

## CASE STUDY: SMART KALASATAMA

Energy saving is becoming part of a broader tendency in urban planning to promote sustainable lifestyles. The emerging activism by cities in energy and climate is reflected in the Smart Kalasatama case. In order to boost new sustainable urban solutions, the Helsinki City Council decided in 2013 to make one of the new construction sites, the Kalasatama harbor area, a model district of smart city development. By 2030 the area will house about 25,000 residents and offer jobs for 8000 people. The process was initiated by a consortium including the local energy company Helen and other large companies to develop new 'smart grid' business. Later, the City of Helsinki placed Forum Virum Helsinki, an innovation intermediary, in charge of the project and Kalasatama was turned into 'smart city' area with more diverse aims (Matschoss and Heiskanen 2017, 2018). The aim is also to create a city district co-designed with citizens with the slogan 'to save one hour of residents' time per day'. The idea is that Kalasatama is a real-life testbed for new services to be scaled up elsewhere. This is done by providing a platform to co-create smart urban infrastructures and services.

Smart Kalasatama is based on the utilization of different technologies and solutions that all use ICT and open data. Several hundred participants—large and small companies, research, public sector, and citizens—are already involved in developing Kalasatama as a smart district. Helen, together with partner organizations, develops smart grid technologies and services such as an electric car network and battery energy storage. The focus is on experimenting with new solutions at varying scales in real life with residents (Mustonen et al. 2017). The Developers' Club gathers city administration, resident associations and businesses in the area together four times a year to discuss the development of Kalasatama. This way of working represents a novel way to cooperate at the city district level in Finland.

The Smart Kalasatama case shows how energy considerations are increasingly embedded in wider urban planning targets. And on the other hand, urban planning—at its best—is not seen merely as physical infrastructure planning. It is also about a redistribution of power, where conventional ways of infrastructure development are challenged, new networks among diverse players are forged, and new solutions are sought for via experimentation. On the other hand, in such a diverse 'smart city' context, energy and resource conservation may have to compete with other agendas, such as the development of new technology and commercial services. In this sense, Smart Kalasatama is a typical case of such developments, with the same strengths and weaknesses. For example, there might be a need for more assessment of whether 'smart' solutions—or new technical solutions in general—deliver the promised environmental benefits, as well as the issue of the extent to which they are scalable (Heiskanen et al. 2017).

An important policy implication from the Finnish cases in general, and the Smart Kalasatama case as an illustration, is that local governments are close to citizens and can influence many of the conditions for sustainable consumption and quality of life. However, local governments might lack the resources, expertise and also the power to innovate, to evaluate projects, and in particular, to scale up innovative practices. Because of this, central governments and the EU might offer more funding for such innovative projects, but also require more and better evaluation and diffusion.

REFERENCES

Annala, S., Lukkarinen, J., Primmer, E., Honkapuro, S., Ollikka, K., Sunila, K., et al. (2018). Regulation as an enabler of demand response in electricity markets and power systems. *Journal of Cleaner Production, 195,* 1139–1148.

Apajalahti, E.-L. (2018). *Large energy companies in transition—From gatekeepers to bridge builders.* Aalto University publication series Doctoral Dissertations, 112/2018.

Haukkala, T. (2018). A struggle for change—The formation of a green-transition advocacy coalition in Finland. *Environmental Innovation and Societal Transitions, 27,* 146–156.

Heiskanen, E., Hyvönen, K., Laakso, S., Laitila, P., Matschoss, K., & Mikkonen, I. (2017). Adoption and use of low-carbon technologies: Lessons from 100 Finnish pilot studies, field experiments and demonstrations. *Sustainability, 9*(5), 847.

Heiskanen, E., Jalas, M., Rinkinen, J., & Tainio, P. (2015). The local community as a "low-carbon lab": Promises and perils. *Environmental Innovation and Societal Transitions, 14,* 149–164.

Heiskanen, E., Lovio, R., & Jalas, M. (2011). Path creation for sustainable consumption: Promoting alternative heating systems in Finland. *Journal of Cleaner Production, 19*(16), 1892–1900.

Hyysalo, S., Juntunen, J. K., & Freeman, S. (2013). Internet forums and the rise of the inventive energy user. *Science & Technology Studies, 26*(1), 25–51.

Järventausta, P., Repo, S., Trygg, P., Rautiainen, A., Mutanen, A., Lummi, K., et al. (2015). *Kysynnän jousto - Suomeen soveltuvat käytännön ratkaisut ja vaikutukset verkkoyhtiöille* [Demand response—Practical solutions applicable to Finland and their impacts on distribution network providers]: *Loppuraportti.* Tampere: Tampereen teknilllinen yliopisto.

Karhunmaa, K. (2018). Attaining carbon neutrality in Finnish parliamentary and city council debates. *Futures* (in Press, Correct Proof). https://doi.org/10.1016/j.futures.2018.10.009.

Karjalainen, S. (2009). Thermal comfort and use of thermostats in Finnish homes and offices. *Building and Environment, 44*(6), 1237–1245.

Laakso, S., Berg, A., & Annala, M. (2017). Dynamics of experimental governance: A meta-study of functions and uses of climate governance experiments. *Journal of Cleaner Production, 169,* 8–16.

Matschoss, K., & Heiskanen, E. (2017). Making it experimental in several ways: The work of intermediaries in raising the ambition level in local climate initiatives. *Journal of Cleaner Production, 169,* 85–93.

Matschoss, K., & Heiskanen, E. (2018). Innovation intermediary challenging the energy incumbent: Enactment of local socio-technical transition pathways by destabilisation of regime rules. *Technology Analysis & Strategic Management*, *30*(12), 1455–1469.

Matschoss, K., Heiskanen, E., Atanasiu, B., & Kranzl, L. (2013). Energy renovations of EU multifamily buildings: Do current policies target the real problems. Rethink, renew, restart. In *Proceedings of the eceee 2013 Summer Study* (pp. 1485–1496). Eceee.

Mickwitz, P., Hildén, M., Seppälä, J., & Melanen, M. (2011). Sustainability through system transformation: lessons from Finnish efforts. *Journal of Cleaner Production*, *19*(16), 1779–1787.

MoEE. (2017). *Government report on the National Energy and Climate Strategy for 2030*. Publications of the Ministry of Economic Affairs and Employment 4/2017. Online: http://urn.fi/URN:ISBN:978-952-327-199-9.

Mustonen, V., Mazur, C., Mattila, M., Hubmann, G., Huuska, P., Jarkko, M., et al. (2017). *Deliverable Proof—Reports resulting from the finalisation of a project task, work package, project stage, project as a whole—EIT-BP16*. Online: http://fiksukalasatama.fi/wp-content/uploads/2017/04/Helsinki-District-Challenge-1_-1.pdf.

NEEAP. (2017). *Finland's National Energy Efficiency Action Plan NEEAP-4*. Report to the European Commission pursuant to Article 24(2) of the Energy Efficiency Directive (2012/27/EU). Online: https://ec.europa.eu/energy/sites/ener/files/documents/fi_neeap_2017_en.pdf.

Rinne, S., Auvinen, K., Reda, F., Ruggiero, S., & Temmes, A. (2018, November 28). *Clean district heating—How can it work?* (Smart Energy Transition Discussion Paper). Online: http://www.smartenergytransition.fi/wp-content/uploads/2018/11/Clean-DHC-discussion-paper_SET_2018.pdf.

Runsten, S., Berninger, K. Heljo, J., Sorvali, J., Kasanen, P., Vihola, J., et al. (2015). *Pienituloisen omistusasujan energiaköyhyys* [Energy poverty among low-income detached house dwellers]. Ympäristöministeriön raportteja 6/2015. Helsinki: Ministry of Environment.

Statistics Finland. (2018a). *Appendix table 1. Energy consumption in households 2010–2017, GWh*. Online: https://www.stat.fi/til/asen/2017/asen_2017_2018-11-22_tau_001_en.html.

Statistics Finland. (2018b). *Final energy consumption by sector by year and sector*. Online: http://pxnet2.stat.fi/PXWeb/pxweb/en/StatFin/StatFin__ene__ehk/statfin_ehk_pxt_010.px/table/tableViewLayout2/?rxid=cfb16ff0-c8b9-4629-b0f3-04af3d02a009.

Statistics Finland. (2018c). *Use of renewable energy continued growing in 2017*. Online: https://www.stat.fi/til/ehk/2017/04/ehk_2017_04_2018-03-28_tie_001_en.html.

Statistics Finland. (2019). *Buildings by area, year of construction, informa-tion and intended use of building.* Online: http://pxnet2.stat.fi/PXWeb/pxweb/en/StatFin/StatFin__asu__rakke/statfin_rakke_pxt_116g.px/table/tableViewLayout2/?rxid=74aac6df-fd8a-48da-86d6-ed04e7fbc49f.

SULPU. (2019, January 15). *Lämpöpumppumyynti ponnahti 22% - Investoinnit yli puoli miljardia* [Heat pump sales increased by 22%—Investments more than half a billion]. News release by SULPU, Finnish Heat Pump Association. Online: https://www.sulpu.fi/uutiset/-/asset_publisher/WD1ExS3CMra3/content/lampopumppumyynti-ponnahti-22-investoinnit-yli-puoli-miljardia?.

CHAPTER 13

# Conclusion: Comparing Household Energy Use Across Europe—Uncovering Opportunities for Sustainable Transformation

*Patrick Naef, Marlyne Sahakian*
*and Gary Goggins*

**Abstract** This chapter considers the similarities and differences between ten European countries in relation to meso-level considerations when it comes to household energy usage. We uncover the governing frameworks and policies related to energy usage, then examine socio-demographic characteristics including housing tenure and location. Next, we consider

P. Naef (✉) · M. Sahakian
University of Geneva, Institute of Sociological Research,
Geneva, Switzerland
e-mail: patrick.naef@unige.ch

M. Sahakian
e-mail: Marlyne.Sahakian@unige.ch

G. Goggins
School of Geography and Archaeology and Ryan Institute,
National University of Ireland Galway, Galway, Ireland
e-mail: gary.goggins@nuigalway.ie

© The Author(s) 2019                                    137
F. Fahy et al. (eds.), *Energy Demand Challenges in Europe*,
https://doi.org/10.1007/978-3-030-20339-9_13

the energy mix and material arrangements, such as building stocks, before turning to climatic considerations and the cost of energy. The conclusion highlights the importance of embedding energy usage in socio-material systems, tackling questions related to collective conventions, for example, as well as notions of sufficiency. While the policy and technological dimensions of energy distribution are easier to account for in country reviews, the collective conventions that hold together everyday practices that use energy services would merit further study.

**Keywords** Socio-material systems · Governing frameworks · Sustainable energy · Sufficiency · Europe

## INTRODUCTION

Households across Europe are key actors in energy transitions towards reduced and improved energy usage, either through the introduction of more efficient technologies in the home, or through more transformative forms of change such as engagement in cooperative renewable energy production. This edited collection presents a series of case studies, which allow for a 'zooming in' around ten, country-specific framings on how energy transitions are being addressed. In this chapter, we 'zoom out' and consider similarities and differences between countries in relation to meso-level considerations. First, we focus on the governing frameworks and policies related to energy usage, then we examine socio-demographic characteristics including housing tenure and location. Next, we consider the energy mix and material arrangements, such as building stocks, before turning to climatic considerations and the cost of energy. The conclusion presents a summary discussion around how these different elements of practices inter-relate, while also noting the lack of comparable data on collective conventions across the countries—an important but often neglected dimension of household energy usage.

## GOVERNING FRAMEWORKS AND POLICIES

In 2016, the average of equivalent tonnes of $CO_2$ emissions per capita in the European Union was 8.7. The countries represented in this collection are dispersed in a relatively balanced way around this value: **Hungary** (6.3), **Switzerland** (6.4), the **United Kingdom** (7.9), **Bulgaria** (8.4) and **Slovenia** (8.6) see their consumption below the European average, whereas **Denmark** (9.3), **Finland** (11.1), **Germany** (11.4), the **Netherlands** (12.2) and **Ireland** (13.5) are above (EEA 2016). Nevertheless, all countries have signed the Paris Agreement based among other factors on a drastic decrease of greenhouse gas emissions, which varies based on country size and energy resources, highlighting the different challenges that these countries are facing towards decarbonisation, and more broadly, towards climate change and pollution mitigation. While there is a common policy agenda around general decarbonisation, there is less consensus around specifics such as nuclear phase out, with opposition movements promoting the increase of nuclear capacity. Some countries, such as **Germany** and **Switzerland**, are engaging in policies away from nuclear, while others, such as **Hungary, Finland** and the **United Kingdom**, still consider atomic energy as an important part of their energy portfolio.

Energy efficiency is now promoted in all national policies and different ways to achieve efficiencies are put forward and implemented with varying degrees of effectiveness. The material presented in previous chapters of this collection show similarities and differences in the policies of the various states under study. For example, policy documents in **Hungary** stress the necessity to improve energy efficiency with a strong focus on the household sector and the building stock, but effective policy support has been volatile. Similarly, there is a strong rhetoric supporting energy efficiency in **Finland**, but actual measures have been relatively limited until recent years. Retrofitting buildings is a widely shared approach to enhance energy efficiency and lower carbon emissions across Europe, but the nature of renewables used between the countries is far from homogenous—which is understandable, given the different natural resources and historical developments. Depending on the climate and the resources available (oil, gas, wood, peat, etc.), different energy portfolios and interests are at stake. There is also an important divide between countries opting for a nuclear phase-out and those in opposition who are working on the expansion of their nuclear capacity. These differences in energy production and management have led to the formation of 'energy islands', with the European Union expressing concerns about the so-called 'fragmentation' of energy policy (Genus and Iskandarova 2018).

The European Union have highlighted some unilateral measures taken by member states, which affect the prices of energy and threaten internal markets. There is thus a call for integration, materialised by the

European Energy Union, of a strategy made up of various dimensions including the reinforcement of energy efficiency to reduce the European Union dependency on energy imports, and climate actions to decarbonise the economy (European Commission 2018). The Energy Union is being developed in five domains of European Union Energy Policy: (i) security of supply (in 2015, the EU28 imported 54% of energy supplied); (ii) sustainability (in 2015 fossil fuels contributed 75% of the fuel mix of EU energy supplied); (iii) greenhouse gas emissions (which for the EU in 2015 was 22% less than the equivalent measure in 1990); (iv) the role of renewable energy in energy supply and use (accounting for 16% of final energy consumption for the EU in 2015); and (v) competitiveness of the EU in the energy sector (European Commission 2017, cited in: Genus and Iskandarova 2018: 11).

Energy security is also a key policy issue shaping political interest in energy supply. Here, Coutard and Shove (2018) bring a more nuanced approach to the question of supply and demand, arguing that unlimited and reliable energy has enabled the normalisation of various forms of energy-greedy consumption, from the use of washing machines and refrigerators, to constant ICT connectivity and the increasing use of air-conditioning in certain contexts. Thus, policy framings around energy could also consider 'How much of what is enough?' (Spengler 2016), or offer **a more explicit focus on sufficiency**. Debates around sufficiency lead into more fundamental and societal questions, such as what services should be enabled, in what contexts, and towards what needs. Taking these debates to heart, and drawing on findings from the H2020 ENERGISE[1] project, Sahakian et al. (2019) propose a definition of sufficiency which accounts for absolute reductions in resource use, while also challenging collective conventions around household energy use, as well as setting upper (and lower) limits to consumption. This approach is built on the premise that while efficiency is a desired approach towards energy transitions, sufficiency should be considered as the first step. Despite the efforts of projects such as ENERGISE, bringing sufficiency debates and ideas into mainstream political and societal discourse remains a considerable challenge.

[1] See www.energise-project.eu.

## Socio-Demographic Characteristics

Socio-demographic characteristics are an important dimension for understanding current energy-related practices and opportunities for change. The usual social categories 'age-gender-(social)class' have effects on how such practices are performed. For example, Gram-Hanssen and Georg (2017) show the link between a person's economic resource, the type of building they live in and their energy habits. Sahakian (2018) also demonstrates the same link in the context of elite households. Economic differences between people lead to different technological acquisition and different habitat conditions, in short, different patterns of consumption. These differences lead several authors to state that thermal conditions should fit the type of building as well as its occupants (Kunkel et al. 2015; Nicol and Wilson 2011; Bopp 2007; Boerstra et al. 2015). The type of occupants and the length of their stay are factors that are classifiable; in other words, they allow standard calculations to apprehend the variability of consumption behaviour. However, as Boerstra et al.'s (2015) critical analysis suggest, these general classifications do not necessarily fit people's perception of comfort and they may lead to higher energy use.

While the links between education levels and energy usage are not linear nor necessarily causal, education levels as a form of social capital may contribute to environmental awareness around energy issues. Age is also seen to have an influence on energy practices, as can be illustrated by the example of laundry. Costanza et al. (2014) suggest that younger people tend to be less predictable in their behaviour and wash at irregular frequencies. Moreover, age has an impact on washing temperatures, as Laitala et al.'s (2012) paper on Norwegian practices indicates. Indeed, they found that younger respondents had lower than average temperatures, and older people had more embedded habits related to washing at hot temperatures.

The location of households, between rural and urban contexts, has a significant influence in the shaping of energy use. It will for instance influence the type and size of the dwelling (e.g. people are more likely to live in a detached house in a rural area), sources of available energy (e.g. district heating may not be an option for rural dwellers) or the connection to the energy system (e.g. the size of a grid can depend on the size of the municipality). Definition of 'urban' and 'rural' varies between countries in Europe, while some countries such as Switzerland and the

Netherlands also have an intermediate category. The OECD proposes a standardised typology based on three categories: (1) predominantly urban; (2) intermediate; (3) predominantly rural (OECD 2019). With the exception of **Slovenia**, where more than half of all people live in rural areas, the countries represented in the previous chapters are mostly composed of urban or intermediate populations, with the **Netherlands**, the **United Kingdom** and **Switzerland** associated with the smallest rural populations. These urban–rural divisions can influence household energy demand and related policies.

Dwelling tenure will also determine households' opportunities to reduce energy demand. Previous research highlights the extensive literature on the **tenant–landlord relations** associated to energy use problematics, pointing out dilemmas such as 'the fact that landlords lack incentives to invest in energy renovations for buildings where the benefits would accrue to tenants or, from the perspective of the tenant, the savings in energy use cannot offset the rent increase due to the renovation' (Laakso and Heiskanen 2017: 12). In this context, **Switzerland** is the only country under study where the share of rental buildings is higher than that of ownership (Fig. 13.1). Of the countries studied here, **Hungary** has the highest share of population living in owner-occupied dwellings (86%), although they are some way behind Romania (96%) as the leading country in the EU.

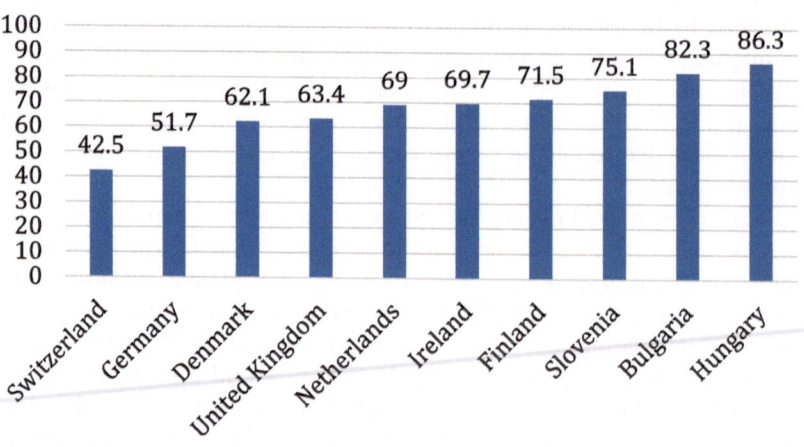

**Fig. 13.1**   Share of owned dwellings (%) (*Source* Eurostat [2018a])

## ENERGY MIX AND MATERIAL ARRANGEMENTS

There is significant variation in the energy mix in the residential sector across Europe (Fig. 13.2). Petroleum products are used mainly by **Ireland** (38%), **Switzerland** (34%) and **Germany** (22%), while they are rarely used in **Bulgaria** (2%). Countries with the highest proportion of renewable energy usage are **Slovenia** (45%) and **Bulgaria** (33%), however high levels of domestic wood burning in these countries has led to problems of poor air quality. The **Netherlands** has only a 5% share of renewables, however the country projects to increase to 12% by 2020; but the target of 17% by 2023 may be more realistic. The **United Kingdom**, with a current share of 4%, has announced that it would provide at least 15% of its energy from renewable energy sources by 2020. Gas is predominantly used by the **Netherlands** (71%), the **United Kingdom** (62%) and **Hungary** (54%), while electrical energy is a resource dominant in **Bulgaria** (42%) and **Finland** (37%), and mobilised in a rather equilibrated way among the remaining countries under study, from 19% (**Hungary**) to 29% (**Switzerland**).

Regarding house characteristics, the type, age and state of dwellings will determine energy use, especially in terms of investment needed to increase energy efficiency. Such characteristics are linked to social trends,

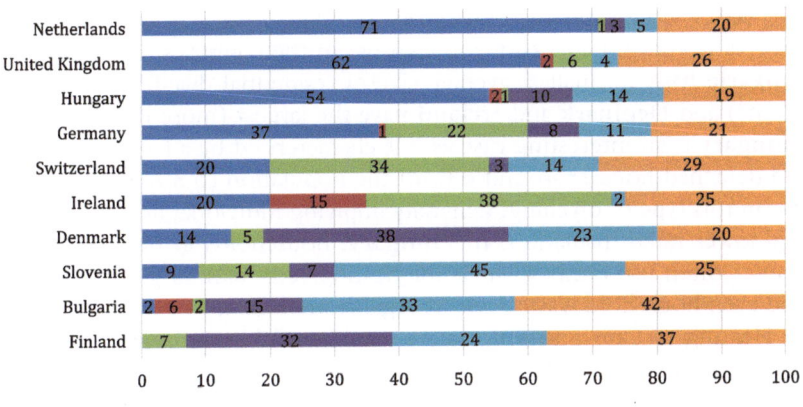

**Fig. 13.2** Share of fuels in the final energy consumption in the resident sector, 2015 (%) (*Source* Eurostat [2018b])

for example towards ever bigger homes and larger window areas. The materials used for the construction of houses have also changed over time, with concrete replacing more traditional materials such as wood and clay. Moreover, the historical development of countries and towns also has an influence on energy use. The necessity for renovation is for instance greater in countries where the building stock is older, such as in **Bulgaria** where only 5% of homes were built after 2000. In a comparison of different countries in Europe, Bartiaux et al. (2014) demonstrate how energy-related renovations did not form a unified practice, but rather a bundle of somewhat disjointed practices. In old and poorly maintained buildings, the practices that are likely to save energy can be quite different from those in new and highly automated buildings. For example, the **United Kingdom** housing stock is one of the oldest compared to most other European countries. Many houses date from the Victorian era and have poor insulation, implying additional energy consumption to maintain a certain level of comfort. However, as the older housing stock is gradually being replaced with a newer one, more energy efficient homes are being developed. In the UK, houses built prior to 1918 represented 25% of the housing stock in 1970, compared to 16% in 2014.

The size and type of the dwelling are also important factors determining energy use (Fig. 13.3). On one hand, larger dwellings consume more energy, but on another hand, multiple rooms offer the opportunity for regulating temperatures when rooms are not used. The average dwelling size varies significantly across Europe, ranging from less than 60 m² in Estonia to more than 120 m² in Cyprus. In the countries reviewed here, **Bulgaria** has the smallest average size of residential dwellings (less than 70 m²) and **Denmark** and **Ireland** have the largest (more than 100 m²). **Hungary** is an interesting case as it is characterised by a large amount of old detached houses: around 63% of the population (6.5 million people) live in this type of dwelling, generally implying individual heating systems and lower rated insulation. In **Hungary**, households living in detached houses often use a mix of fuels for heating (e.g. natural gas and wood) and even household waste (despite legal restrictions).

Heating systems and sources are also material conditions influencing household practices related to energy use, with great differences across Europe. Indeed, various sources of energy for heating are predominantly used depending on the countries. One issue with calculating the share of fuels in final energy consumption has to do with definitions of energy sources, and the temporality of the assessment. For example, what

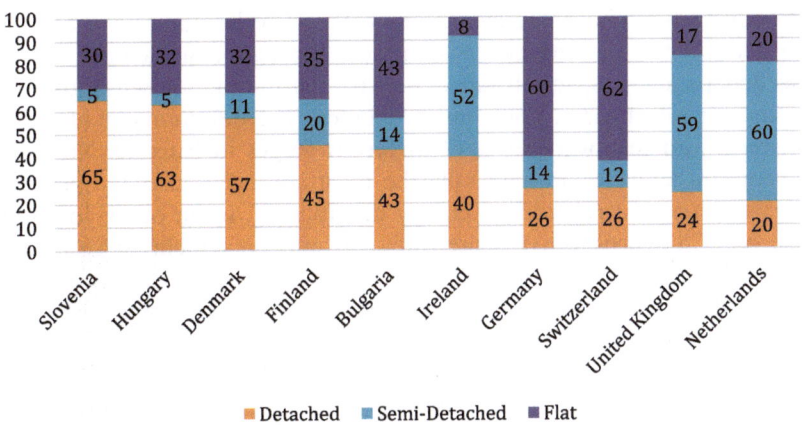

**Fig. 13.3** Common dwelling types, 2015 (%) (*Source* Eurostat [2017])

is considered as a 'renewable' resource in one context may not be in another (e.g. hydraulic). In the **Netherlands** and the **United Kingdom**, households heat space in a large majority with natural gas (87% for the **Netherlands** and 76% for the **United Kingdom**). This is also the case, but with a lesser extent, for **Hungary** (49%) and **Germany** (44%). Renewable energies are the dominant fuel for space heating in **Slovenia** (63%) and **Bulgaria** (58%), with **Hungary** using 40% renewables (note that wood is currently classified as a renewable form of energy). Petroleum products for space heating are used in a very contrasted manner depending on the countries, ranging from almost half of the fuels for **Switzerland** (46%) and **Ireland** (46%), to being almost zero in **Bulgaria**, the **Netherlands** and **Hungary**. Derived heat is used mostly in **Denmark** (38%) and **Finland** (35%), but very scarcely in other countries, from 13% (**Switzerland**) to 0% (**United Kingdom**). Finally, solid fuels and electrical energy are the least used, the former being used most significantly by **Ireland** (23%) and the latter by **Finland** (25%).

In recent years, the development of '**smart systems**' and '**smart cities**' has increased to become a global phenomenon. Smart technologies have been progressively integrated in government agendas and considered as a priority area for research. Moreover, smart meters are starting to be installed in homes all over the world towards the goal of improving household energy efficiency. Smart technologies can be used to act directly on energy consumption through management of the needed parameters, for instance a smart thermostat for self-regulating homes or self-managing washing machines. The development of **smart systems and technologies is at different stages in Europe**. Northern European countries are generally the most advanced, already operating smart systems in relation to energy distribution, while other countries are only in a prospective stage, building and implementing strategies. Since 2010, smart energy systems gained momentum in **Denmark** and **Finland**, with the latter providing most of its electricity consumers with automatic meter reading installations. Both countries invested significantly in smart grid and smart energy research. **Finland** is already developing smart products (e.g. Internet of things, building automation, smart controls) for the export market and has around 20 cities piloting smart technologies. In the **United Kingdom** and the **Netherlands**, national governments committed to ensuring smart meters for all households by 2020, and smart metering is also part of the 2050 Energy Strategy of **Switzerland**. **Germany** is significantly upgrading its electricity grid, integrating smart technologies, and developing the concept of smart cities in major urban areas such as Berlin, Munich, Mannheim and Hamburg.

The number and the ownership of energy suppliers, from private to state-owned, are also elements that may influence household practices in relation to energy usage. State regulations related to energy distribution generally seek to protect consumers' interests and act primarily on the energy bill (adapting the cost of energy between supply and demand for example), which can also shape household energy-usage. For example, regulations to offer low electricity prices based on renewable energy sources might shape the consumer environmental sensitivity. The scale of the production and distribution (municipal, regional, national) of energy will influence the number of suppliers (from one national supplier, to multiple regional or local ones). For example, the Dutch state owned 'Transmission System Operator', owns and operates the high voltage transmission grid at the national level in the **Netherlands**, but other parts of the grid (lower voltages until 230–400V) are owned and operated by regional energy companies. **Switzerland's** and **Finland's** energy systems are also mostly public, but they are operated at the level of municipalities, hence decision-making is decentralised and arguably more responsive to local contextual considerations.

## CLIMATIC CONDITIONS AND THE COST OF FUELS

Climatic conditions will also have an influence on how energy-using practices play out, especially in terms of heating and cooling, but also in relation to other practices such as the option for drying clothes outdoors or the necessity to refrigerate foodstuff. The categorisation of different climates varies according to the sources and the description methods, and moreover, climates are far from being homogenous in the whole national territory. In **Switzerland**, the temperature significantly depends on altitude, with high variation from Arctic to Mediterranean types of climate. Furthermore, climate change will have a significant impact on the Swiss climate: local climatic warming in the Alpine arc is twice as important as the global average. **Finland** has an annual average temperature of 2 °C (a little more than 5 °C in Helsinki on the South Coast and about 0 °C in Sodankylä in Lapland). Furthermore, this Nordic country is associated with a great variability in the availability of sunlight over the year: during winter solstice, the sun is up for less than 6 hours in Helsinki, and during summer solstice, for almost 19 hours. Thus, with a focus on countries situated in Central and Northern Europe, the main implication for household energy use is related to home heating.

The price of energy in Europe depends on a mix of conditions, including climatic conditions as mentioned above, as well as access to energy sources, and levels of subsidies and taxation. In 2017, the average electricity price for household consumers was 0.20 (Euro per kWh) in the European Union (0.21 in the European area). The average price of natural gas for EU household consumers was 0.06 Euro per kWh (0.07 in the European area), with a range of between 0.04 and 0.11 Euro per kWh. The lowest prices are in **Bulgaria** and in **Hungary** (the highest, in Sweden), where the respective governments have an explicit aim to keep household energy prices low, and where access to affordable energy remains an important concern for a large proportion of the population. The **cost of electricity** may also vary depending on the type of consumer (e.g. household, business, industry), the time of consumption (where day or night rates apply) and the type of the band (kWh capacity). In **Germany** and **Denmark**, where systems are mostly operated by private energy companies, the prices are the highest at more than 0.30 Euro per kWh. In contrast, electricity prices in Bulgaria are among the lowest in Europe at less than 0.10 Euro per kWh.

If we compare the prices of some of the energy sources mentioned above (natural gas, electricity), **Bulgaria** and **Hungary** have among the lowest range of prices in Europe, which seems logical since these countries are also the ones with the lowest GDP per capita. Yet, besides this constant, there are some significant differences between the rankings of countries, depending on the resource in question. These variations are of course related to the living standards of these various states, but also to the availability of the resource within their boundaries or the price it costs to import it. For instance, **Switzerland** is characterised by high prices when it comes to diesel or natural gas—since there are no gas and oil sources in the country—however, the prices of electricity are in the lower range, a situation that could be explained by the important hydroelectric resources present in this mountainous region. In contrast, electricity costs more in **Germany** than almost anywhere else in Europe, a situation associated with the country's attempt to transition from fossil fuels and nuclear energy to more renewable energy sources. This transition is importantly funded by high taxes on energy companies, as in the case of **Denmark**, which is also among the most expensive countries in terms of household electricity.

Fuel subsidies also have a role to play on the cost of energy, as do social subsidies indirectly—in that they can provide support for people in need. Framing household's energy practices by suggesting that sustainable behaviours can be financially beneficial is seen to increase acceptance and adoption of sustainable consumption practices. For example, **Denmark** had a tradition of offering various subsidies, for instance for the installation of solar panels and the replacement of oil burners with heat pumps. Moreover, many subsidies were available in buildings construction and renovation for thermal insulation and double-glazing. However, recent policy changes tend to remove subsidies to households and concentrate on energy savings in business, implying cuts in subsidies to home-renovations.

Several specific campaigns targeting household energy practices have also been promoted this last decade, often based on financial mechanisms. The **German** government's largest current campaign is the '*Deutschland macht's effizient*' initiative focusing primarily on energy efficiency by offering information, consultations and financial incentives in the form of grant aid for households, companies and municipalities who undertake to improve their energy efficiency. In the **Netherlands**, the 'energy efficiency you do now' programme provides cheap loans for energy efficiency renovations to private home owners and associations of apartment owners.

The price of energy can influence consumers but is not sufficient in itself to explain energy usage. The cost of energy must be placed in relation to revenues and other household expenditures. Many European countries including the **UK**, **Hungary** and **Bulgaria** are facing particularly important challenges in terms of price and access to energy; it is estimated that fuel poverty affects over 15% of British households (approximately 4 million) and 21% of the Hungarian population (Fülöp and Lehoczki-Krsjak 2014). Over 40% of Bulgarian households are unable to heat their homes to an adequate level. On the contrary, **Switzerland**, with a relatively high buying power and low prices of electricity (compared to healthcare costs for example) uses a significant amount of electric heating. In **Germany**, with higher prices of electricity, a lower amount of expensive electric heating is used.

## Summary and Discussion

In comparing and contrasting energy-related problem framings and social, material and institutional make-up across Europe, this concluding chapter highlights that there is no one-size-fits-all solution to the energy challenge and that policies for energy demand reduction have to carefully consider and address the differences in material and institutional constitutions of energy demand and energy systems, locally, regionally, nationally and cross-nationally. The summaries presented by each country in this edited collection give some interesting insights on what trends are currently underway, and what this implies for the future of energy demand in Europe. The initiatives that focus on household energy reveal interesting findings, such as the importance of EU funding as well as other national funding schemes towards promoting innovative approaches to reduced energy usage, as well as the significance of working with multiple stakeholders towards community engagement, which is seen as preferable to national-led or 'top-down' initiatives. In this vein, there seems to be increasing interest—in policy discourse if not in action—on the need to move away from the 'passive consumer' to the 'active citizen' when it comes to framing the role of households in energy transitions. While most of the energy initiatives seem to be focused on individual and technological change, there are promising examples of how initiatives can also address more complex representations of change. Here, it could be relevant to consider not only the question of access to energy—with energy poverty a key issue in several countries, from the UK to Bulgaria—but also upper limits to energy usage, in some contexts and in relation to certain practices.

Another main conclusion is the need to embed energy demand in socio-material systems, tackling questions related to cultural context— for example, the dominance of 'car culture' in the German case. While the policy and technological dimensions of energy distribution are easier to account for in country reviews, the collective conventions that hold together everyday practices that use energy services are relatively under studied. This relates to a key question in social science approaches to energy: 'how do conventions around energy services evolve, how do they alter over time, and how can they be changed once they are cemented?' (Sovacool 2014: 19). Comparable data is available on energy-related policies, forms of energy provisioning, technological configurations, climatic factors or socio-demographic aspects, to name but a few angles, but there is a lack of comparable empirical data on the collective conventions around energy use and across European countries. The ENERGISE project contributes to this research gap by producing new learnings on social conventions around energy usage across eight countries in Europe, with a focus on heating and laundry, and providing insights on how upper and relative limits to consumption can lead to reflections on 'how much is enough?' (For example, see Sahakian et al. 2019), thus contributing to energy transitions in specific contexts and cultures. In addition, the good practice examples of sustainable energy initiatives provided in this book can serve as inspiration to policy-makers, practitioners, businesses, NGOs, students, academics and others interested in creating a more sustainable and just future.

## References

Bartiaux, F., Gram-Hanssen, K., Fonseca, P., Ozoliņa, L., & Christensen, T. H. (2014). A practice–theory approach to homeowners' energy retrofits in four European areas. *Building Research & Information, 42*(4), 525–538.

Boerstra, A. C., van Hoof, J., & van Weele, A. M. (2015). A new hybrid thermal comfort guideline for the Netherlands: Background and development. *Architectural Science Review, 58*(1), 24–34.

Bopp, K.-F. (2007). *Housing, energy and thermal comfort: A review of 10 countries within the WHO European Region.* Copenhagen: WHO Regional Office for Europe.

Costanza, E., Fischer, J. E, Colley, J. A, Rodde, T., Ramchurn, S. D., & Jennings, N. R. (2014). Doing the laundry with agents: A field trial of a future smart energy system in the home. In *Proceedings of the SIGCHI Conference on Human Factors in Computing Systems, CHI '14* (pp. 813–822). New York, NY: ACM.

Coutard, O., & Shove, E. (2018). Infrastructures, practices and the dynamics of demand. In E. Shove & E. Trentmann (Eds.), *Infrastructures in practice: The dynamics of demand in networked societies* (pp. 10–22). Oxon: Routledge.

EEA. (2016). *Greenhouse gas emissions per capita.* Accessed 21 March 2019 from https://ec.europa.eu/eurostat/tgm/table.do?tab=table&init=1&language=en&pcode=t2020_rd300&plugin=1.

European Commission. (2018). *Energy union and climate.* Available at https://ec.europa.eu/commission/priorities/energy-union-and-climate_en. Accessed 13 November 2018.

Eurostat. (2017). *Distribution of population by dwelling type, 2015.* Available at http://ec.europa.eu/eurostat/statistics-explained/index.php?title=File:Distribution_of_population_by_dwelling_type,_2015_(%25_of_population)_YB17.png. Accessed 21 March 2019.

Eurostat. (2018a). *Housing statistics.* Available at https://ec.europa.eu/eurostat/statistics-explained/index.php?title=Housing_statistics. Accessed 21 March 2019.

Eurostat. (2018b). *Energy consumption in households.* Available at https://ec.europa.eu/eurostat/statistics-explained/index.php/Energy_consumption_in_households. Accessed 7 June 2018.

Fülöp, O., & Lehoczki-Krsjak, A. (2014). Energiaszegénység Magyarországon. *Statisztikai Szemle, 92*(8–9), 820–831 (in Hungarian only). Available at http://www.ksh.hu/statszemle_archive/2014/2014_08-09/2014_08-09_820.pdf. Accessed 26 April 2018.

Genus, A., & Iskandarova, M. (2018). *Policy paper 1: State of the art and future of policy integration for EU policy on energy consumption.* ENERGISE—European Network for Research, Good Practice and Innovation for Sustainable Energy, Deliverable No. 6.4.

Gram-Hanssen, K., & Georg, S. (2017). Energy performance gaps: Promises, people, practices. *Building Research & Information, 46,* 1–9.

Kunkel, S., Kontonasiou, E., Arcipowska, A. Mariottini, F., & Bogdan, A. (2015). *Indoor air quality, thermal comfort and daylight, Buildings Performance Institute Europe (BPIE).* Accessed 21 March 2019. http://bpie.eu/publication/indoor-air-quality-thermal-comfort-and-daylight-an-analysis-of-residential-building-regulations-in-8-member-states-2015/.

Laakso, S., & Heiskanen, E. (2017). *Good practice report: Capturing cross-cultural interventions.* ENERGISE—European Network for Research, Good Practice and Innovation for Sustainable Energy, Grant Agreement No. 727642, Deliverable No. 3.1.

Laitala, K., Klepp, I. G., & Boks, C. (2012). Changing laundry habits in Norway. *International Journal of Consumer Studies, 36*(2), 228–237.

Nicol, J. F., & Wilson, M. (2011). A critique of European Standard EN 15251: Strengths, weaknesses and lessons for future standard. *Building Research & Information, 39*, 18–193.

OECD. (2019). *National area distribution (indicator).* Accessed 21 March 2019 from https://doi.org/10.1787/34f4ec4a-en.

Sahakian, M. (2018). Constructing normality through material and social lock-in: The dynamics of energy consumption among Geneva's more affluent households. In H. Allison, D. Rosie, & W. Gordon (Eds.), *Demanding energy: Space, time and change* (pp. 51–71). Cham: Springer International Publishing.

Sahakian, M., Naef, P., Jensen, C., Goggins, G., & Fahy, F. (2019). Challenging conventions towards energy sufficiency: Ruptures in laundry and heating routines in Europe. *ECEEE Summer Study 2019 Proceedings.*

Sovacool, B. K. (2014). What are we doing here? Analyzing fifteen years of energy scholarship and proposing a social science research agenda. *Energy Research & Social Science, 1*, 1–29.

Spengler, L. (2016). Two types of 'enough': Sufficiency as minimum and maximum. *Environmental Politics 25*(5), 921–940.

# Index

154    INDEX